Historic Iron and Steel Bridges
in Maine, New Hampshire and Vermont

ALSO BY GLENN A. KNOBLOCK
AND FROM McFARLAND

Black Submarines in the United States Navy, 1940–1975 (2011)

*African American World War II Casualties and
Decorations in the Navy, Coast Guard and Merchant* (2009)

*"Strong and Brave Fellows": New Hampshire's Black Soldiers and
Sailors of the American Revolution* (2003)

Historic Iron and Steel Bridges in Maine, New Hampshire and Vermont

Glenn A. Knoblock

McFarland & Company, Inc., Publishers
Jefferson, North Carolina, and London

LIBRARY OF CONGRESS CATALOGUING-IN-PUBLICATION DATA

Knoblock, Glenn A.
Historic iron and steel bridges in Maine,
New Hampshire and Vermont / Glenn A. Knoblock.
    p.   cm.
Includes bibliographical references and index.

ISBN 978-0-7864-4843-2
softcover : acid free paper ∞

1. Iron and steel bridges—Maine.
2. Iron and steel bridges—New Hampshire.
3. Iron and steel bridges—Vermont.
4. Historic bridges—Maine.
5. Historic bridges—New Hampshire.
6. Historic bridges—Vermont.
I. Title.

TG24.M2K59 2012   624.2'170974—dc23   2011051551

BRITISH LIBRARY CATALOGUING DATA ARE AVAILABLE

© 2012 Glenn A. Knoblock. All rights reserved

*No part of this book may be reproduced or transmitted in any form
or by any means, electronic or mechanical, including photocopying
or recording, or by any information storage and retrieval system,
without permission in writing from the publisher.*

Front cover image: a portal view of the Highgate Falls Bridge,
Highgate, Vermont (photograph from the author's collection)

Manufactured in the United States of America

*McFarland & Company, Inc., Publishers
Box 611, Jefferson, North Carolina 28640
www.mcfarlandpub.com*

This book is dedicated to my mom,
an intrepid adventurer,
and my dad, an engineer.
Thank you for all your support over the years!

Table of Contents

Acknowledgments   ix
Introduction   1

PART I: HISTORY AND DEVELOPMENT OF NORTHERN NEW ENGLAND BRIDGES

1. Northern New England's Early Bridges   5
2. The Advent of Iron and Steel Truss Bridge Designs   10
3. The Anatomy of a Bridge   19
4. Notable Bridge Types of Northern New England   31
5. The Life of a Bridge   47
6. Civil Engineering in Northern New England   55
7. Railroad Bridges   64
8. The Death of a Bridge   72
9. Preservation Efforts   78

PART II: NOTABLE BRIDGE HISTORIES

10. Vermont Bridges   89
11. New Hampshire Bridges   110
12. Maine Bridges   140
13. Interstate Bridges   167

Appendix 1: Bridge Companies Represented in the Region   187
Appendix 2: Vermont Historic Metal Highway Bridges   188
Appendix 3: New Hampshire Historic Metal Highway Bridges   191
Appendix 4: Maine Historic Metal Bridges   194
Bibliography   197
Index   203

Acknowledgments

With all the recent attention and focus on the condition of our national infrastructure, coupled with the discussion and debate surrounding President Obama's Stimulus Plan, it is perhaps inevitable that a related book like this should be written. However, the true "stimulus" for this book came in the form of a phone call from Maggie Stier of the New Hampshire Preservation Alliance. With metal truss bridges added to their list of "Seven to Save" structures in 2008, she was seeking help to get the plight of these bridges recognized by a wider audience. With my previous book on New Hampshire's covered bridges and the lectures I've given on them statewide, I have somewhat of a sympathetic bridge audience and a forum from which to speak. Indeed, while writing and researching my covered bridge book, I had gathered quite a bit of information on old iron and steel bridges, but had never thought about doing anything with it. Maggie's call, with additional encouragement from state historian Jim Garvin, got me hooked and eventually led to the book you now hold in your hand. Originally starting with a focus on my home state of New Hampshire, my efforts soon expanded to our northern New England neighbors.

As usual, any book of this nature requires the help of a great many people. First and foremost, I owe a great debt to four individuals, all bridge experts in their respective states; James Garvin, in addition to his encouragement, was vital to my efforts in gathering records for New Hampshire bridges, and patiently answered my many requests and queries, both via email and in person. His help was even more valuable given his prominence and years of experience as a New Hampshire historian and author. The same is true of Professor Robert McCullough of the University of Vermont at Burlington, the head of the VAOT's Historic Bridge Program. In addition to his helpful emails and patiently answering my many questions and requests for information, he made my visit to Montpelier to view state bridge records both easy and fruitful. His book on Vermont bridges is an outstanding work, about the only such published work of its kind at present in New England. It is thoroughly researched, well written, and was a book on which I heavily relied. Last, but not least, are Kirk Mohney of the Maine Historic Preservation Commission and Kurt Jergenson, formerly historic planner in the Environmental Office of the Maine DOT. Kirk was extremely helpful during my trip to MHPC headquarters in Augusta and patiently answered my many questions. Kurt was invaluable in providing me with updated bridge lists for Maine, copies of the survey reports on all pre–1950 metal truss bridges in the state, and patiently answered my questions about current bridges and their current condition and future plans for bridge removals.

No less important to getting this book researched were the following individuals who went above and beyond; Matthew Thomas, friend, historian extraordinaire, and my able navigator during a trip to visit the Maine bridges; Paul Hallett of the Conway Scenic Railroad, who kindly allowed me to accompany him on a maintenance run from Conway to Fabyan Station. Though I surely sounded like an investigative reporter with all the questions I asked of him, Paul had all the answers I was seeking, and more. Indeed, of all my bridge outings, this one was surely the most thrilling; Tony Keegan, a trainman who put me in touch with

Paul Hallett and answered my many railroad-related questions. Of course, this book could not have even come close to being a reality without the support, encouragement, and patience of Terry, my wife and best friend for over thirty years now, and my daughter Anna. They not only put up with my absences during the course of my field research (about 2,500 miles by my reckoning), but also endured my bridge talk at home, the accumulation of mounds of research materials, and the occasional "detour" while on a family trip to check out a nearby bridge.

The following individuals, in no particular order, were also helpful in providing information and photos about the bridges in their town, or their area of expertise. While all of the information they provided or the bridges in their respective towns may not have made it into the book, the insights I gained from this wealth of material were key in developing an overall understanding of how metal bridge building progressed in the region: David Ruell (Ashland, NH), Ron Reed (Boscawen, NH), Walt Stockwell (Campton, NH), Anne Dunbar (Limington, ME), Suzy and Dave Schaub (Union, ME), Brian Lombard (Railroad Operating Engineer, NH Bureau of Rail and Transit), Alan King Sloan (King Iron Bridge Co.), Jim Hodsdon (Ashland, NH), Duane Lewis (Bridge Operator, Southport, ME), Michael Baldock (United Kingdom photographer), Gordon Page (Vice President, Director of Passenger Operations, Maine Eastern Railroad), Hank Pokigo (Director of Marketing, Hardesty & Hanover, LLP), Martha Turner (Richmond, VT), Katie Reilley (New Haven, VT Highway Dept.), Craig Hanchey (Bridgehunter.com photographer), John Davis (Grand Trunk Railway New England Lines historian), Charles Nye (Highgate, VT), Lucille Keegan (Bristol, NH), Christine Bannerman (Wallingford, VT), Julie Moore (Wilmington, VT), Terri Parks (Dalton, NH), Barbara Holt (Franconia, NH), Stephen Perkins (Bennington Museum, VT), Barbara Mace and Eleanor Porrill (Goffstown, NH), Ryan Parent (New England rail photographer), Jim McElroy (Henniker, NH), Marie Stanley (Hill, NH), Charles Hammer (Monroe, NH), Sally Kriebel (Thornton, NH), Paul Hudson (Bethlehem), Euclid Farnham (Tunbridge, VT), Janet Hayward-Burnham (Bethel, VT), Bill Reed (New Sharon, ME), Ray and Hildy Danforth (Shelburne, NH), Mrs. Bill Paradis (Stratford, NH), John Buxton (Bridge Maintenance Engineer, Maine DOT), Bill Troup (Secretary, Sandy River & Rangeley Lakes Railroad, ME).

While the list above is largely complete, I may have inadvertently missed mentioning a few people and apologize for doing so. There were also a number of people, local residents mostly, whom I met during my bridge expeditions throughout Maine, New Hampshire, and Vermont that offered help, directions, and information. While your names may not be listed, your help was equally valuable and, if you're reading this, you most likely know who you are! Despite all the assistance I have received from all these individuals, named and unnamed, in the end any errors that may have crept into this work are purely of my own doing. All photographs are by the author, or from my collection unless otherwise noted.

Introduction

Bridges, in all their varied forms, real or imagined, are a major part of our lives, both physical and spiritual. Indeed, is there a day that doesn't goes by, perhaps, where we encounter a bridge of one kind or another? Consider, if you will, the following popular metaphors:

"Don't burn your bridges behind you"
"We'll cross that bridge when we come to it"
"That's water under the bridge"
"The bridge to nowhere"

Not a fan of metaphors? Then, how about the following the literary allusions, touching on the spiritual, the humorous, and the practical:

"Faith is the pierless bridge supporting what we see unto the scene that we do not" (Emily Dickinson, American poet).

"We build too many walls and not enough bridges" (Isaac Newton, English scientist).

"I would rather be the man who bought the Brooklyn Bridge than the man who sold it" (Will Rogers, American humorist).

"The grave is but a covered bridge leading from light to light through a brief darkness" (Henry Wadsworth Longfellow, American poet).

"There is nothing in machinery, there is nothing in embankments and railways and iron bridges and engineering devices to oblige them to be ugly. Ugliness is the measure of imperfection" (H. G. Wells, English author).

Ok, still not convinced? How about the bridge in popular literature and film? Here is a brief sampling:

The Bridge of San Luis Rey (Thornton Wilder, American novelist, 1927)
The Bridge Over the River Kwai (Pierre Boule, French novelist, 1952)
The Bridges of Toko Ri (James Michener, American novelist, 1953)
The Bridge on the Drina (Ivo Andric, Yugoslavian novelist, 1945)
A Bridge Too Far (Cornelius Ryan, Irish American historian, 1974)
The Bridges of Madison County (Robert James Waller, American novelist, 1992)
The Bridge to Terabithia (Katherine Paterson, American author, 1977)

And, finally, what about the bridge in music? Here are a few popular songs you may have heard over the years:

London Bridge Is Falling Down (popular nursery rhyme)
Bridge Over Troubled Waters (Paul Simon/Art Garfunkel)
59th Street Bridge Song (Feeling Groovy) (Simon and Garfunkel)
Brooklyn Bridge (Frank Sinatra)
The Bridge (Neil Young)
London Bridge (Fergie)

The point of all this popular culture stuff is simple; bridges, in all their varied forms, beautiful or ugly, are an important part of most everyone's daily life. Yet how often do we stop to contemplate the real bridges in our lives? Thus, we get to the point of this book in hand, the historic iron and steel bridges of northern New England. For over one hundred and thirty years now they have performed the jobs for which they were designed and built. Celebrated in their early years for their advanced technology, many of them are now perceived as nothing more than rusted relics of a bygone era. Some, though their days are numbered, still remain on busy thoroughfares, taken for granted by harried motorists in the commuting age. Most lie on scenic back roads and byways, often bypassed, while others lie abandoned and forlorn on roads long since closed, the day when they might collapse looming in the not too distant future.

Maine, New Hampshire and Vermont combined once had hundreds, perhaps nearly a thousand, of these geometrically appealing highway and railroad bridges, the first generation of scientifically designed spans, crossing their rivers and roadways. Some were regarded as engineering marvels at the height of the Industrial Age, during the period from 1870 to 1900, while others built from 1905 to 1940 were regarded as remarkable examples of civil engineering in the modern age. Despite the fanfare that often accompanied their building, many of these remaining bridges are endangered landmarks. As of this writing, less than 250 historic iron and steel truss and related bridges remain in the region. The loss of many of these bridges seems inevitable unless state or local historical entities intervene on a timely basis. It is for these reasons, both their historical significance and their dwindling numbers, that these bridges are now, finally, being recognized among the public at large for their historical and engineering significance. Indeed, New Hampshire's iron and steel truss bridges are now on the New Hampshire Preservation Alliance's "Seven to Save" list of significantly endangered architectural structures, while Vermont has instituted a highly successful Historic Bridge Program to preserve its historic structures that has served as a model nationwide.

How and why northern New England came to build these new bridge types over the years, notable examples of bridges no longer standing, and an account of those that still remain is the subject of the work in hand. This is a book intended for anyone who has an interest in bridges, whether they are a bridge enthusiast (a "pontist"), civil engineer, native New Englander or someone from outside the region. Since I am a historian, not an engineer, I have kept, I hope, the engineering jargon and technical information to a reasonable amount — enough to inform the general reader, but not so much, either, so as to overwhelm or turn this work into a dry technical account. The majority of this work is concerned with iron and steel truss bridges, though I have also included some additional unique categories of bridge types that also saw major advances with the coming of iron and steel. Since this is not a comprehensive history of bridge building in northern New England, I must remind the reader that there are several types of metal bridges, most notably girder and I-beam bridges, that are not covered in this work. Though structures of this type are extremely important in the history of the development of bridge design, aesthetic appeal in these structures is largely lacking and their technical details are generally above the knowledge of most laymen, myself included.

New England is often noted for its old things, whether it be houses, churches, barns, or graveyards. Whether you use this book as an armchair guide, or whether it ends up as a dog-eared companion in your car's glove compartment, perhaps with a few notes penned in based on your own personal travels, it is my fondest hope that after reading it you will have acquired in your mind's eye yet another image to add to the list of "old things" considered quintessentially "New England."

Part I

History and Development of Northern New England Bridges

1

Northern New England's Early Bridges

Before there were iron and steel bridges in the region, the wooden bridge was king. By briefly considering the development of earlier bridge types, a greater understanding of their metal successors will thereby be gained. Timber was not only cheap and readily available in this heavily forested region, but their builders from early times were skilled at putting these materials to good use in spanning local rivers, streams, and brooks. Stone arch bridges were also constructed before the advent of iron and steel truss bridges; but the early use of this type of bridge was not nearly as widespread due to both the skill required in their building by experienced masons as well as the inherent transportation difficulties entailed in getting quarried stone to more distant locations before the advent of rail and motorized transportation. By 1920 stone bridges in northern New England were largely abandoned in favor of concrete or steel bridges, except in some areas (such as Barton, Vermont, and Milford, New Hampshire) where granite was locally quarried and available as a building material. However, some attractive stone arch bridges were also built early in the 1900s for aesthetic reasons in the spirit of the Good Roads movement in such New Hampshire towns as Rochester, Hillsborough and Merrimack, as well as Manchester, Vermont, to name just a few. Ironically, because of the durability and lower maintenance costs of a stone bridge, those built in the aforementioned New Hampshire towns sometimes replaced early iron bridges that were too lightly built to carry the loads necessitated by modern transportation needs.

The first wooden "bridges" were simple affairs indeed, often consisting of nothing more than a fallen tree across a river or stream on which precariously balanced travelers could pass. Such a crossing might suffice for a time, but as settlements grew into towns in 18th and 19th century, other, more dignified and proper means of crossing were soon constructed. In Harrisville, New Hampshire, a fallen pine tree was the only method of crossing Goose Brook to reach the lower part of town before George Jones built a 120-foot long wood bridge adjacent to his own property as a replacement for the fallen pine in the 1840s. The simplest of the early wooden bridges were pile and trestle bridges, built in large numbers on tidal streams, small rivers, and marshy areas in the New Hampshire Seacoast and Maine coastal areas. One of the earliest documented types of this bridge in Maine, though certainly not the first, was Sewall's Bridge in York, Maine, built in 1761. This bridge, a simple structure of a type dating back to ancient times, was easily constructed and maintained, and easy to replace if damaged. Another early type of wooden bridge utilized in coastal regions was the boom bridge, a span consisting of logs floating on the water, with a hewn surface face up. Further inland, where annual spring flooding was common, small bridges were built using the king-post or queen-post trusses—standard wooden bridge designs that had been used in Europe for centuries and ones with which most any builder was likely well acquainted. Even today, these simple type of wooden truss bridges are commonly constructed by homeowners in the region and elsewhere who have small streams that need to be crossed, whether on foot for decorative purposes, or by vehicular traffic. Like the pile and beam bridge, the king-post and queen-post truss bridges could be easily replaced if washed out by a spring freshet, as often happened. Utilizing simple truss designs, these earliest wooden truss bridges are but little documented and seldom exceeded 40 feet in length. More efficient and stronger bridging solutions would need to be developed as New England expanded west and northward and commerce grew apace.

The apex of the art of open wooden bridge construction in the region was reached within a twenty-year period, begun by a builder, Timothy Palmer of Newburyport, Massachusetts, renowned in both New England and the mid–Atlantic seaboard. It was continued by local builders, most utilizing an arch design. In New Hampshire, impressive advances were made with the construction of the Piscataqua Bridge across the Piscataqua River from Fox Point in Newington to Meader's Neck in Durham. Completed in November 1794, this bridge was engineered by Timothy Palmer and was the largest structure of its kind in the United States at the time. Indeed, the bridge was often simply referred to as the "Great Arch Bridge." Its cost has been variously stated to have been between $62,000 and $66,000. It measured some 2,362 feet long, nearly a half-mile, and 38 feet wide, consisting of pile and beam spans for most of its length from either end with a 244-foot long arch over the main shipping channel of the fast-flowing river between Goat and Rock Islands. Built as a toll bridge by local merchants, the construction materials consisted of three thousand tons of pine timber, eighty thousand four-inch planks, twenty tons of iron, and eight thousand tons of stone. Towering some 60 feet over the water at its height and painted a gleaming white, the Piscataqua Bridge was a modern engineering marvel that cost foot travelers an extravagant, for the times, 2½ cents to cross, horse and rider 8 cents, a sulky 20 cents, a sleigh 12 cents, and a four-wheeled cart or carriage 40 cents. This significant bridge, whose western terminus was at the site of the beginning of the First New Hampshire Turnpike, required extensive maintenance over the years. A lottery was established in 1804 to raise $15,000 for repairs, with tickets selling for $5 apiece. Subsequently repaired, the bridge needed further attention in 1830 when part of it gave away, and again in 1854. After this the bridge was sold to private interests for just $2,000, a substantial sum to be sure, but nothing compared to its original cost. This decrease in value was due to the advent of rail travel on the Seacoast which lessened the importance of the bridge. When 600 feet of the bridge was destroyed by ice during the winter of 1855, the remainder of the bridge was removed because the owner of the bridge could not afford the repair costs.

With the great success of the Piscataqua Bridge, builders to the west on the Merrimack and Connecticut rivers in New Hampshire and Vermont were inspired to build substantial spans of their own, though not always with satisfactory results. A single arched bridge measuring 236 feet long, modeled after Palmer's arch, was constructed by Rufus Graves, a local builder, across the Connecticut River between Hanover, New Hampshire, and Norwich, Vermont, in 1796. But the structure crashed into the river due to its own sheer weight just eight years later. Moody Bedell, a well-known builder from Haverhill, New Hampshire, constructed a 1,000-foot long bridge across the Connecticut River to Newbury, Vermont, in 1795 and 1796 that consisted of four arches and one draw span. This bridge served travelers until 1818 when it was carried away by ice, the sad fate of many a bridge in New England down through the years. Other noted open bridges in the region include the 1797 Kennebec River Bridge at Augusta (when Maine was still a district of Massachusetts), also built by Timothy Palmer, and those designed by Vermont bridge builder John Johnson (a native of New Hampshire), including a four-span king-post (also called a "rafter truss") truss bridge on the Winooski River between Colchester and Burlington contracted for in 1816.

While the open-arch bridges built or inspired by Timothy Palmer were impressive examples of the bridge builder's art, it was the wooden covered bridge, utilizing a wide variety of trusses often in combination with the wooden arch that would soon become the predominant bridge type by the second decade of the 19th century in New England and the mid–Atlantic states. The first known covered bridge in America was built, albeit grudgingly, by none other than Timothy Palmer across the Schuylkill River in Philadelphia in 1806. The idea of constructing a roof to shelter the bridge trusses from the elements and thus prolong the life of the optimistically named Permanent Bridge was conceived by a member of the bridge company that contracted Palmer for the work. This novel, yet simple idea was so radical for the time that even a builder as renowned as Palmer had to be persuaded to cover his impressive open trusses. It took some time for the covered truss bridge to catch on in more rural northern New England, though perhaps the first covered bridges in the region were erected with little fanfare on a small scale and have gone undocumented. It would not be unreasonable to speculate that the first covered bridges in the region were small in size, perhaps a simple kingpost truss in Bennington County, Vermont. However, despite such conjecture, the honors for northern New England's first covered bridge must go to the covered bridge spanning the Kennebec River at Augusta. This bridge was built in 1818 to replace Timothy Palmers' bridge on the same site by Benjamin Brown and Ephrain Ballard, Jr., and utilized Palmer's tried and true arch truss design. Vermont's earliest known covered bridge was the Burr arch truss bridge across the Mississquoi River at Highgate Falls. Built in 1824, it was replaced in 1887 by an iron lenticular truss bridge still in existence today. New Hampshire's first documented covered bridge is the Woodsville Bridge. Still standing today, this Town lattice truss bridge was built across the Ammonoosuc River between Bath and Haverhill from 1827 to 1829 and cost the hefty sum of $2,400.

Though covered bridges were constructed almost entirely from wood until they went out of favor, their development over the years in truss design and construction is very relevant to the iron and steel bridges that came to supplant them. Not only did these seemingly quaint bridges pave the way for more modern iron and steel structures, but their construction also served as a proving ground of sorts for America's first civil engineers. Covered bridge truss designs from the early days of Timothy Palmer relied on the use of arches for long spans that required great skill to construct, while smaller crossings, those well under 100 feet in length, were limited to the simple king and queen-post structures. However, Ithiel Town of Connecticut would forever change the landscape of bridge building in America. Trained as an architect by the noted Boston builder Asher Benjamin, Town established his own firm in New Haven, Connecticut, in 1810 and in 1820 patented his soon to become famous Town lattice truss. The beauty of this design (truly resembling a modern day garden lattice) lay both in its simplicity (any carpenter could build it) and economy of design, using common-sized lumber that was readily available and utilizing but few metal components. Moreover, this truss for the first time allowed the building of straight (as opposed to arched) bridges in much longer spans than those previously constructed. Interestingly, Town proposed early on that his truss design could also be built of cast or wrought iron, but this would not become a reality until 1859.

One truss designer, whose name is all but forgotten except by students of engineering history, is Col. Stephen H. Long, a native of Hopkinton, New Hampshire. A noted engineer with the U.S. Army, Long patented his own "assisted truss" in 1830, consisting of a row of panels, each resembling a boxed X, augmented by a king-post truss above the center panels. Of Long's design, one historian states "that he knew a valid method for calculating stresses in the member of a truss," but "does not give any details of such analysis" (Condit, pg. 93). Though Long's achievements seem to have been well appreciated by European engineers, they were but little heralded in the U.S., where competition among bridge designers was fierce. Stephen Long deserves to be better remembered today, for the concept of calculating stress would become, and still is today, the fundamental basis for iron and steel bridge building.

Long's design would be improved upon within ten years when William Howe, a native of Massachusetts, patented a modified lattice truss design that introduced the use of wrought iron rods as tension members. Though Howe's original patent design was not widely used, a modified version designed by Amasa Stone in 1841 was essentially the same as Stephen Long's truss design, with the exception that the upright posts in each of the boxed X panels, except for the end posts, were made of wrought iron. Though this design, universally referred to as a Howe truss, was largely based on the truss originally designed by Stephen Long, his attempt to win a patent infringement case was unsuccessful. In the last half of the 19th century, the Howe truss design was very popular and countless bridges of this type were built, many of them of the railroad variety. The age of iron, albeit on a small scale, had begun in American bridge building. Now, it was only a matter of time before the era of the wooden bridge was over.

Along with the change in truss design over the years in covered bridge construction, the capabilities of builders on the local and regional level also necessarily evolved. Covered bridge builders of the past are typically viewed today as country craftsmen who built bridges where there was an immediate need that had to be met. The structures they built were sometimes simple, not always "square" and seldom fancy, but they were sturdy and have stood the test of time. This is the romantic view of the country builder, but this idea holds water only to a very limited extent. True, some of the region's covered bridge builders *were* simple craftsmen; James Tasker built a number of sturdy multiple king-post bridges in the Cornish, New Hampshire, area and was largely illiterate, but he also built at least two bridges across the Connecticut River with Bela Fletcher (no small achievement), while a youthful Walter Piper, just 18 years old, built a covered bridge in Lyme, New Hampshire. However, these men, accomplished though they were, are only a small part of the story. Many covered bridge builders were also accomplished engineers and a number of builders were acclaimed well beyond their home state borders. Among the Vermont men that fall into this category are Nicholas Powers, who built the still extant covered bridge at North Blenheim, New York, an impressive double-barreled bridge, and many others in his home state as well as bridges as far distant as Maryland; the previously mentioned John Johnson, who built a two-span Burr arch-truss bridge at Essex, Vermont, in 1825; and Farewell Weatherby, the colorfully named builder of an impressive double-barreled Burr arch-truss bridge over the Lamoille River at Cambridge in 1845. New Hampshire had its own builders of note, including "Boston" John Clark of Franklin, who built the Republican Bridge in 1839, an unusual hipped-roof structure, and possibly several railroad bridges in the area; Peter Paddleford of Littleton, who designed his own variant of the Long truss that saw extensive use in Maine, New Hampshire, and as far away as Ohio; the Childs brothers of Henniker, Horace, Enoch, and Warren, who formed their own bridge building company and patented their own truss design, built the sophisticated wooden McCallum truss structure (the Rainbow Bridge) across the Merrimack River at Boscawen, and later gained renown for their railroad bridge building activities in Maine; and John C. Briggs, a civil engineer of Concord who patented his own wooden truss design and built the triple railroad bridges ("the Triplets") at a difficult spot on the Merrimack River at Hooksett. While the covered bridge builders from the state of Maine did not gain the same renown as those from New Hampshire or Vermont, that does not mean that they were any less skilled. Among the more active bridge builders in Maine were Colonel Thomas Lancaster of New Sharon, a builder of several double-barrel bridges in the 1830s in Franklin County, and his successor Robinson Davis. Another prominent covered bridge builder active in the region was Captain Isaac Damon of Northampton, Massachusetts. A partner of Ithiel Town, Damon not only built a lattice truss bridge across the Penobscot River, but also others throughout New England and New York, including twelve such bridges across the Connecticut River.

Skilled though these covered bridge builders were, the dawn of the Railroad era and the gradual introduction of motorized transport spelled the eventual end of large-scale wooden bridge building. But, it was not just these factors that brought about the use of iron, and later steel; it was the sweeping changes in American society and science that made the use of iron and steel inevitable in the science of bridge building.

2

The Advent of Iron and Steel Truss Bridge Designs

The change from wooden to iron bridge building was very slow in coming in America, gradually evolving over decades. The first iron bridge ever built was that over the River Severn at Coalbrookdale in England, completed in 1779. Interestingly, the famed American Revolutionary and pamphleteer, Thomas Paine (the author of "Common Sense" and "The Rights of Man"), made models of a cast-iron arch bridge in 1786 and even had components for the bridge cast in England. However, despite his best efforts, Paine failed to interest anyone in its building. The first use of iron in American bridges was utilized by the Pennsylvanian James Finley, who first built one of his "chain" bridges in 1801 and was granted a patent for his design in 1808. Finley's chain bridges were suspension bridges and about forty were built in the United States, the most prominent being the Newburyport Bridge across the Merrimack River in Massachusetts, close to the New Hampshire border. Another patent was issued to August Canfield of New Jersey in 1833, but this too was essentially a suspension bridge and did not see wide use. In 1836 the first cast iron arch bridge was built in America at Brownsville, Pennsylvania, across Dunlap's Creek. Built by Captain Richard Delafield of the Army Corps of Engineers, it replaced a Finley chain bridge. However, it would take many more years of experimentation and truss development before the iron bridge gained wide acceptance for larger spans. Timber was still less expensive and old traditions died hard. Some early well-publicized cast iron bridge collapses both in the United States (at Ashtabula, Ohio, in 1876) and abroad (the Firth of Tay Bridge in Scotland in 1879) served to slow the acceptance of iron bridges for larger spans until further advances in technology.

The first iron bridge type to come into common usage in the eastern United States was the bowstring arch truss patented by Squire Whipple in 1841. Rightfully known as the father of American iron bridge building, Whipple was born in Massachusetts in 1804, moved at an early age to New York and was later educated at Union College. By the 1830s Whipple was a surveyor for the Baltimore and Ohio Railroad and later worked for the Erie Canal before joining the New York and Erie Railroad as its chief engineer. After patenting his bowstring truss, so named for its resemblance to an archer's bow, Whipple built his first example of this bridge—its top and bottom chords made of cast iron and the vertical and diagonal members from wrought iron—in New York in 1842. Just four years later, Whipple patented his own trapezoidal truss for railroad bridges that gained wide acceptance and use; and in 1847 published his landmark tome, *A Work on Bridge Building*, the first work of its kind to discuss the distribution of stress in bridge components and one that served as a reliable handbook for bridge construction. Though undocumented, it is highly likely that Whipple's bowstring truss bridge was the first iron bridge type to be built in northern New England, probably in the 1860s.

While Squire Whipple was the "father" of American iron bridge design, the first "scientifically designed truss" (Condit, pg. 109) is often credited to the Boston bridge builder,

Thomas Pratt. Born in 1812, the son of noted Boston architect Caleb Pratt, Thomas began working in his father's office at age 12, and at age 14 attended Rensselaer Polytechnic Institute in Troy, New York. Graduating from RPI with an engineering degree while still a teen, the younger Pratt began a long career working for various New England railroads. He patented his first truss in 1842; this was a modified version of Amasa Stone's Howe truss utilizing wooden chords and posts and wrought iron diagonal members. Two years later he and his father were granted a joint patent on a similar design, with either parallel chords or a polygonal top chord that could be built of wood and iron, or entirely of iron. The Pratt truss first came into wide use in railroad bridge construction in the 1850s, though wood was still the most trusted material for longer spans until the 1870s. Though Squire Whipple's railroad truss was more popular than the Pratt truss, the later would achieve a greater endurance. It become a standard in highway bridge building, and many bridges of this type were built of iron and steel in the region as late as the 1940s. Just as with wooden truss designs, there were modifications to such designs as the Pratt truss that would become bridge standards, giving rise to new designations. One popular variant of the Pratt truss is the Parker truss, named after the noted Boston engineer Charles H. Parker. After the Pratt's patent had expired, he modified their elongated polygonal top chord into a bridge truss type often referred to as "camelback" or "humpback" bridges.

With the greater use of iron in American bridges, a large number of trusses designed by American engineers came to be patented: some of them, as will be discussed in detail further on, were quite revolutionary and long-lived; others, such as the lenticular (or parabolic) truss would have a short life; while yet others, such as the Truesdell truss, would prove untrustworthy and dangerous for all but the smallest spans.

A number of factors contributed heavily to the development and use of iron and, later, steel bridges in northern New England and nationwide by the end of the 19th century. Most of these coincided with developments that evolved throughout America to varying degrees, influenced further by local conditions.

Advances in Metallurgy

The production of metal and the science of metallurgy, practiced in America since the colonial days, advanced at a break-neck pace during the great age of the Industrial Revolution. The first type of iron produced in America was cast iron. While this type of iron had been produced in Europe for hundreds of years and was cheap to produce in America, it contained many impurities that resulted in a low tensile strength. Heated to a high temperature, the iron was then poured into molds of different varieties and lengths produced in local foundries to fashion bridge components beginning in the 1830s. Though useful in some early bridge truss applications (one advantageous property is its compressive strength) cast iron under heavy loads and other stresses was brittle and easily prone to sudden failure if used in the wrong application. While the earliest iron bridge in America, the previously mentioned structure over Dunlap's Creek in Brownsville, Pennsylvania, was made of cast iron and has had a long life, this and other bridges made of cast iron into the 1860s were all arch bridges that used the compressive strength of the material to its inherent advantages. However, in more advanced truss bridges it was unsuitable.

Wrought iron was the next form of iron to be used in bridge building in America. This type of iron had been first used in America in the decade before the Revolutionary War, but was expensive to produce and, before the advent of large-scale manufacturing and the development of rolling mills after the 1850s required skilled ironworkers. Interestingly, in form wrought iron precedes that of the liquid state of molten cast iron and was produced at a lower temperature, the iron being continually worked ("wrought") by hand (and later rolled by machine) until hardening. Though requiring a great amount of labor to produce, wrought iron was very pure, giving the finished product a high tensile strength, good elasticity, and a high degree of corrosion resistance, all extremely desirable qualities in bridge building materials. Indeed, it was the introduction of the wrought iron members in otherwise all wooden bridge designs, such as the Howe truss, that later came to spell the end of wide-scale wooden bridge building. The first all iron bridge truss built in New England in great numbers, Whipple's bowstring arch truss, used a combination of both cast and wrought iron components. However, with increased mechanized production of wrought iron, cast iron quickly disappeared from widespread use after 1870. It was from this time, until the 1890s, that wrought iron reigned supreme in bridge building.

Iron production was important in early metal bridge building, but its heyday lasted less than fifty years, when it was gradually replaced by the use of steel, beginning with the construction of the Eads Bridge in St. Louis (1868 to 1874) and the famed Brooklyn Bridge in New York (1869 to 1883). Steel, a complex material made from a large variety of ferrous metals, was produced prior to this time, but was extremely expensive and thus was not used in buildings and bridges until after 1856 with the introduction of the Bessemer process in America. This ground-breaking method of refining steel, however, was limited in capacity, and it would not be until other processes were developed and refined that steel became readily affordable after 1875. In property, early steel beams resembled those made of wrought iron, but were superior in nearly all areas to a degree of some ten or fifteen percent. The only drawback to steel was (and is) its higher risk of corrosion than iron, making it necessary to keep steel structures, especially older bridges, regularly painted to avoid exposure to the elements. This inherent defect in the use of steel continues to be a source of difficulty in the preservation of historic steel bridges, most notably in the case of the vertical lift Memorial Bridge over the Piscataqua River between Portsmouth, New Hampshire, and Kittery, Maine. Steel production technology would also see great advances in the 20th century, and modern steel bridges are constructed with a special weather resistant type of steel that does not require painting.

The Rise of the Railroads

The great expansion of the railroad network in New England and beyond after the 1850s was another factor that accelerated the development and use of metal truss bridges. The overall impact on America due to the advent of the railroad has been well discussed and need not be recounted in detail here, except to say that by the 1890s all but the smallest of communities in the Northeast were connected to the outside world by means of rail. With the rapid increase of the rail network, there quickly arose the need in northern New England to build crossings over many brooks, creeks, streams, and a great many rivers of all sizes, from the wandering Walloomsac and the meandering Lamoille rivers in Vermont, the mountain-born Ammonoosuc and the swiftly flowing Contoocook and Souhegan rivers in New Hampshire, as well as the most northern St. John and the coastal Royal rivers in downeast Maine. However many rivers the region has, when it came to bridge building, the most important waterways were those that have always loomed large, literally and figuratively, in the region. These include the mighty Merrimack River in New Hampshire, the majestic Connecticut River bordering Vermont and New Hampshire, the Winooski River that slices through north-central Vermont, the swift flowing Piscataqua River that borders Maine and New Hampshire, and the long and swift-flowing waters of the Kennebec and Penobscot rivers in Maine.

No matter what size the crossing, the railroads charged forward, and soon came to favor metal bridges over those made of wood. Too, the growing use of electric streetcar systems, or trolleys, in a number of towns and cities by the late 1800s also served to hasten the change from wooden to metal bridges. Covered bridges especially were inadequate structures to facilitate electric streetcars for one simple reason; the very fact that they *were* covered meant that such bridges could not easily accommodate the overhead lines that carried electricity to power this new mode of travel. Towns as widely divergent in size as Nashua and Goffstown, New Hampshire, serve as interesting examples of how communities were forced to deal with changing methods of transportation over the years and the challenges they posed to town planning officials. In the early 1880s, Nashua replaced the Taylor's Falls covered bridge with a spectacular three-span lenticular truss iron bridge, while in 1901 Goffstown replaced its beautiful single-span Town-lattice covered bridge on Main Street over the Piscataquog River with a low (pony) Pratt truss bridge. Though many communities took pride in their modern bridges, these structures did not always live up to their expectations and not everyone was enthralled with their ascetics. The new Taylor's Falls Bridge in Nashua, despite the fact that it was made of iron, was not nearly strong enough to handle the large volume of electric streetcar traffic and had a relatively short life before being replaced by a more substantial stone bridge in the early 1900s. The new bridge in Goffstown was termed by some an "iron monstrosity." Indeed, back then as it is now, what was termed "progress" by many sometimes depended on the eye of the beholder!

The Rise of American Mills and Manufacturing

Another factor associated with America's Industrial Revolution that promoted the use of iron bridges was the vast expansion of mill activities and their building complexes and infrastructure that seemed to blossom throughout the region. Among these manufacturing towns and cities, producing mostly textile and pulp paper products, to facilitate the building of first generation iron bridges were those in Manchester, Claremont, Dover, Berlin, and Nashua in New Hampshire, and the mills at Yarmouth, Fairfield, Windham, Mechanic Falls, Madison, Lewiston, Auburn, and Bangor, to name just a few, in Maine. Rural Vermont had less of these manufacturing sites, but such commercial cities and towns such as Rutland, Hartford, Highgate, Montpelier, Springfield, and Johnson built iron bridges as well to accommodate the rise of local industry.

Indeed, all of these locales, and a whole host of smaller sites in numerous other towns and cities throughout New England saw their greatest period of growth and prosperity in the late 1800s and were thus well capitalized to expand their facilities to increase production, and they did so by erecting the most modern facilities of the time. This not only included larger buildings and new equipment, but also new iron bridges that would connect mill buildings on both sides of a given river, or one mill building to another, so that workers, raw materials, and finished goods could easily get from one area of a mill's operation to another. While textile production of materials such as wool and cotton and finished materials such as shoes, as well as lumber and such related building products as doors, sashes, furniture, and pulp paper processing comprised the bulk of the industrial activities at these sites, more unusual products included gunpowder, quarried stone, and a wide variety of machinery. Sadly, while many of these iron bridges are undocumented and were long ago torn down, period postcards, surviving photos, and popular lithographs of prominent mill complexes all serve to highlight the existence, importance, and early introduction of iron bridges in these industrial settings. It is not surprising, then, that the earliest known iron truss bridge still surviving in the region is the small bowstring arch bridge spanning the Sugar River in Claremont, New Hampshire, at Monadnock Mills #6. This bridge, built by the Moseley Iron Bridgeworks of Boston in 1870, is an important survivor due not only to its age, but also truss type and its mill setting. How-

The iron arch bridge in Mechanic Falls, Maine, was long gone by 1907. Note the sign designating a $3 fine for going faster than a walk on the bridge!

ever, it must not be thought that all such mill bridges were small structures. One of the most impressive mill bridges built in northern New England was the double-decked lenticular truss bridge built across the Merrimack River at Manchester, New Hampshire, serving both the city and two large mill concerns. Constructed in 1880 and 1881 by the Corrugated Metal Corp. of East Berlin, Connecticut, the McGregor Bridge consisted of three spans, as well as a bowstring arch truss on the western approach and was 930 feet long. The upper deck was open to general traffic, while the lower deck was reserved for mill workers.

The Good Roads Movement

A final major factor that promoted the cause of iron and steel bridge building in northern New England was the advent of the Good Roads movement beginning after 1890. It is rather ironic that this national movement was spurred on not by the rise of the automobile (as many mistakenly believe) or by any great social cause hoping to improve the conditions of rural America, but more simply by the rise of the bicycle. Though various early forms of what is now called the bicycle first appeared overseas in the early 1800s, it was not until the late 1880s that a more modern version became wildly popular in America, one that had a steerable front wheel, chain-driven rear wheel, diamond-shaped frame, and pneumatic tires that were the same size. The bicycle soon caught on with all segments of society and was hugely popular. For men, it became a reliable and easy mode of transport to work in the city, while for women it revolutionized the lives of many and offered them a form of personal freedom to which they had previously been unaccustomed! As a natural outgrowth of this popularity, bicycling clubs were quickly established; their riders not content with just riding the city streets, they soon took to the back roads of America. What they found off the beaten path, however, were far from ideal conditions. For decades, American roads had been derided for their poor conditions and all but the best, usually those located close to town or city, were constantly in a rutted and bumpy state. Riders, whether on bicycle, horseback, or horse-drawn wagon, were

The Advent of Iron and Steel Truss Bridge Designs

The McGregor Bridge in Manchester, New Hampshire, was one of the largest factory bridges in New England and is indicative of the city's status as an important industrial center.

choked with dust in the dry summer, and bogged down in the mud in wet conditions. Indicative of these conditions were the comments made in 1870 by Connecticut governor Marshall Jewell, a native of Winchester, New Hampshire, who said, "What we complain of under the present condition of affairs is that all four wheels of our wagons are often running on different grades. This kind of road will throw a child out of its mother's arms. We let our road-makers shake us enough to the mile to furnish assault and battery cases for a thousand police cases" (Sloane, pg. 58).

While the bicycle craze may have started the Good Roads movement, the Rural Free Delivery Act, passed by Congress in 1893, guaranteeing mail delivery to remote areas of the country, gave the movement added focus and increased the funding for transportation network expansion and improvements over the next several decades. The growing use of the automobile, followed soon thereafter by heavier trucks, would soon led to the formation of the state agencies today responsible for road and bridge construction and maintenance. Thus it is that bridge building, and road construction, activities in northern New England gradually shifted from villages, town, and city to centralized agencies at the state level. Increased help in building America's infrastructure for individual states came with the passage of the Federal Aid and Works Program in 1916. With this important piece of legislation, the Office of Public Roads (changed to the Bureau of Public Roads in 1919, and subsequently the Federal Highway Administration in 1967) in Washington, D.C., contributed matching funds to road and bridge projects nationwide in an effort to improve America's roads. Federal assistance would subsequently reach even greater levels in 1933 with President Franklin D. Roosevelt's New Deal administration.

While the federal government would gain increasing importance in improving road conditions over the years with funding, the first standardized efforts to improve conditions on the local level were made by state level agencies. The establishment of state bridge and highway departments in the region varied slightly between the three states and each offers interesting insight as to how bridges and roads were built and maintained.

New Hampshire

In late 1903 the Good Roads League was established in New Hampshire, with ex-governor Frank W. Rollins, a long-time supporter of state tourism, as president and noted civil engineer and bridge designer John W. Storrs as secretary. The organization's stated goal was simple; "that we may have as good roads in New Hampshire as in any state and by doing so we will be able to compete in the markets with all comers, being able, as we will, to transport the produce of our farms to the railroads and merchandise from the railroads to our stores and farms with the least possible expense" (Metcalf and Norris, pg. 292). While wider and stronger bridges were soon developed with motorized transport in mind, it was also recognized by the men of the Good Roads League that horse travel would remain predominant for some time in rural areas, and that better roads and bridges would also mean less wear and tear on farmer's horse teams. Where multi-horse team wagons may have previously been required to bring products to market over badly rutted roads, better graded and maintained roads would surely reduce the number of horses needed, and thus the farmer's overall expenses.

The Good Roads League was a logical public sector outgrowth in support of the state government's efforts to improve New Hampshire's infrastructure system; that same year, 1903, a system of state roads in the White Mountains was adopted, followed in 1905 by the official formation of the State Highway Department, with a state engineer, the experienced John Storrs, in charge of New Hampshire's roads and bridges. This department, the ancestor of today's New Hampshire Department of Transportation after subsequent expansion and reorganization efforts in 1915 and 1950, was vital in championing the use of modern metal bridges. Just as was the case with New Hampshire's noted wooden covered bridge builders, so too were the state's metal bridge designers and engineers acknowledged nationally for their skill and expertise.

However important the Good Roads movement was on the state level, action at the city, town, and village level, where control of the local infrastructure was maintained for many years was vital. Indeed, the editors of the *Granite State Monthly* in 1904 warned that "the people in New Hampshire can not wake up any too quick for their own good on this matter; a little permanent improvement each year will soon make a good showing, and conditions will become better," and offered the stern proposition that "towns that are careless, negligent, and behind the times are surely going to suffer" (*ibid.*, pg. 393). That every town or region in the state did not, or could not financially, heed these warnings is reflected in the remarks of a travel writer, who, while journeying through the Monadnock region by auto in 1915, commented that "the main roads in the region are good gravel or dirt. The less said about the others the better" (Johnson, pg. 87). By and large, however, New Hampshire communities, large and small, did respond and invested large amounts of money into their infrastructure by purchasing road pavement machinery and replacing wooden bridges with more modern iron structures.

As might be expected, the state capitol city of Concord was in the forefront of these efforts, building its first iron bridge in 1872 and soon thereafter systematically replacing other wooden bridges with those of iron and steel as the need arose. Similar efforts by forward-thinking small communities would follow by the end of the century, as highlighted by the bridge building activities of the village of Campton. Though it would (and has to this date) retain several of its covered bridges, eight metal bridges were constructed in Campton from 1885 to 1911, two of which were built in the year 1899 alone. Such efforts by Campton were duplicated in numerous other towns and cities throughout the state, and these activities combined, by serving to fulfill the goals of the Good Roads League, also served to greatly expand the focus of metal bridge building in the state. By doing so, the era of the covered bridge as the crossing of choice was gradually coming to an end in New Hampshire. Even the last covered bridges built in the state for purely financial, versus tourist or historic considerations, had a close relationship to the iron and steel bridges that would make them obsolete. The Mount Orne Bridge in Lancaster and Columbia Bridge in Columbia were built in 1911 and 1912 (respectively) by the Berlin Construction Company of Connecticut, a firm long known for its iron bridges and used many metal components. The covered bridge at Hancock, constructed in 1937, was designed by a steel bridge engineer employed by the state, Henry Pratt, Jr., and was constructed of Douglas fir timber augmented by three tons of steel (Hancock, pg. 52) by a firm, Hagen and Thibideau, heavily involved in building steel bridges elsewhere in the state.

Maine

The Good Roads Movement was equally active in Maine, where the transportation network in the later half of the 19th century was decidedly inferior. In 1891, the Maine division of the League of American Wheelsmen (the premier bicycling organization nationwide) declared that "correct road building has never been regarded as a subject worthy of investigation" (Maine State Highway Commission, pg. 7), but also recognized that "roads cost money and Maine is not a wealthy state" (*ibid.*). Beginning in 1901, the state government began to fund local highway and bridge projects, and in 1903 spent some $40,000. In 1905, the Commission of Highway Office was established, with Paul D. Sargent, an engineering graduate from the University of Maine, as chief. In 1907 the Maine State Highway Department was created, and in 1913 a Highway Commission, consisting of three members was formed as the state continued to refine its organizations responsible for road and bridge building and maintenance. The passage of the Bridge Act of 1915 established the Bridge Division of the highway department, stating that bridges would be planned and built on a "thoroughly business-like basis and will give the state good substantial bridges properly designed and built" (Maine State Highway Commission, pp. 14–15). Annual appropriations funded the Bridge Division in its first few years of operations, but by 1920 bonds were issued to fund its projects. Bonding continued to fund projects until 1937, when bridge funds were allocated out of Maine's General Highway Fund.

The first two chief engineers of the Bridge Division, Llewellyn Edwards and Max Wilder, are noted for their service in improving the state's highways and bridges. While the metal bridges they caused to be built were not considered innovative in form, neither was there a need for such. In true New England style, these engineers built bridges using solid and well-proven forms. Conventional these bridges in Maine may have been, but that does not mean that they weren't impressive. One of the state's first major bridge projects was the International Bridge between Van Buren and St. Leonards, New Brunswick, a four-span combination truss bridge measuring 451 feet in length over the St. John River which was completed in 1911.

Vermont

Even before the era of metal bridge building, Vermont had a long tradition of supporting the building and maintenance of roads and bridges. To be sure, authority in this area still rested with each town, but the Vermont Constitution of 1777 was progressive enough to include an article addressing the law of eminent domain, as well as fair and just compensation to a property owner whose land was taken for use by the general public. The first highway statutes in the state were soon enacted, including one in 1779 that imposed a direct state highway tax on every male aged 16 to 60, requiring them to work at least four days a year on town roads (and be paid for such), or pay a fine. A subsequent consolidation of highway laws in 1797 made towns liable for any damages caused by poor bridges and roads. Interestingly, while towns remained responsible for building and maintaining bridges, the Vermont legislature later authorized landowners to petition a town for the building of a bridge. An attempt to consolidate road and bridge building activities began in 1827 with the establishment of county road commissioners, five for every Vermont county. These men were appointed yearly and reviewed all requests within their jurisdiction for new roads and bridges, and could order repairs on existing roads and bridges as well. This power, of course, contradicted the traditional power held by local town selectmen, and the county road commissioner office was abolished in 1831. Still, town selectmen in Vermont did not hold all the power; county grand juries were authorized by the state in 1834 to indict towns that failed to maintain their bridges, or even for failure to build a needed bridge! Indeed, bridges were often on the minds of local selectmen in Vermont.... One case from the state that reached the U.S. Supreme Court in 1848 involved the condemnation of the West River Bridge, a private toll bridge, between Brattleboro and Dummerston in 1843. The court ruled against the West River Bridge Company, allowing selectman to condemn the structure and compensate its owners.

The Good Roads Movement came to Vermont in 1892 with the establishment of the Vermont League of Good Roads. That same year, the Board of Highway Commissioners was established, as was a state highway tax to provide funding for improving Vermont's roads. This tax, five cents for each individual on a town's tax list, replaced the old labor tax of 1779, but with one caveat; it could not be used for bridge repair and replacement. Bridges thus remained the domain of local selectmen. The leading figure in these new laws and related activities was a trained engineer and forward thinker, Levi Fuller. Born in New Hampshire and later a resident of Brattleboro, he served as state governor from 1892 to 1894 and was elected as the first president of the Vermont League of Good Roads. While the first Board of Highway Commissioners only lasted four years, it was soon replaced in 1898 by the office of state highway commissioner. Bridges would remain under local control until the state established its Bridge Fund in 1915. Though more rural in nature than most of the rest of New England, Vermont, too, was influenced by the rise of automobile ownership; fewer than 300 iron bridges were likely built in the state prior to 1900, but that number would increase greatly in the first decades of the 20th century. While covered bridge building persisted in the Vermont longer than in any other New England state, again due to its rural nature, the public perception of the covered bridge would, by the end of the 1920s, begin to be transformed from that of an economically viable structure of the present to that of a rural American icon symbolic of days gone by.

3

The Anatomy of a Bridge

Before proceeding to a discussion of the specific bridge types and their use in the region, it will here be helpful to offer a layman's guide to what constitutes a truss, the three general types of truss bridges, as well as terminology (identified in *italic* form) associated with standard bridge features and concepts.

A *truss* bridge, whether made of wood or metal, can be simply defined as a jointed structure which is arranged to carry loads in such a way that each of its members is subject to stress only in the direction of its length and that the loads carried are supported only at the *joints*, the points where the members meet. These joints were typically pinned or riveted during the time under consideration. In a *pinned joint*, the earliest form of connection used, at each panel point a round bar, or pin, passes through a hole in the members being connected and served to help transfer the stress from one member to another, giving the bridge a degree of flexibility. Bridges connected in this manner are categorized by their low or high pin connections. Notable examples of early pin-connected bridges are still extant in the region; those in Vermont include the Vine Street Bridge in Northfield (1870), the Highgate Falls Bridge in Highgate (1887) and the Howard Hill Road Bridge in Cavendish (1890), though this last bridge has been dismantled and is currently in storage. Those in New Hampshire include the Livermore Falls Bridge in Campton (1885), two lenticular truss bridges over the Gale River in Franconia (1889), the Connecticut River bridge between Stratford and Maidstone, Vermont (1893), the abandoned Thompson Crossing Road Bridge connecting Antrim and Bennington (1893), and the now relocated Pingree Bridge in Salisbury (1893).

Riveted joints were typically used in pairs and connected an iron or steel plate to each adjoining member, making a bridge more rigid than those that were pinned. Those who favored this method claimed it was less likely to collapse due to the failure of a single member. Rivets were more expensive and did not come into common usage until after 1910. In fact, both types of connecting systems were employed in New England and throughout America and both, when utilized according to accepted building practices, performed equally well. The small Park Street Bridge in Exeter, New Hampshire, built in 1892, is an early example of a riveted structure, as is the Rice Farm Road Bridge, also built in 1892, in Dummerston, Vermont.

The top member of a bridge truss is termed the *upper chord*, while that at the bottom is termed the *lower chord*. The members that connect the upper and lower chords are called *braces*, *web members*, or *diagonals*. The arrangements of these components into varied, often triangular shaped *panels* determined the truss type (Pratt, Warren, Parker, etc.), while the size of the bridge itself determines how many panels will be utilized. Except in cases where triple trusses were utilized for large bridges or those carrying heavy loads, usually for railroad traffic like the Elm Street Bridge in Biddeford and the St. John Street overpass in Portland, both in Maine, trusses were built in pairs to support a roadway or bridge floor.

Single members in truss panels are usually referred to as *diagonals* due to their position, while added members that results in a truss panel resembling a boxed X are usually referred

Pinned joints, such as this on the Highgate Falls Bridge in Vermont, were the earliest form of bridge connection technology.

to as *counter diagonals*, or sometimes simply as *counters*. The way in which each of these support members acts in distributing the load of the bridge is at the very heart of bridge design, ideally resulting in a structure that is both safe and economical to build manner. Without delving too deeply into the scientific principals of bridge engineering (see the list of sources at the end of this book for technical works that cover this subject in greater detail), two key terms that describe how different truss components work in relation to each other are *tension* and *compression*. Put very simply, a member under tension is pulled upon, while one under compression is squeezed or pressed. When either one of these forces are at work, the bridge is under *stress*. Those trusses, and the bridges they supported, whose stress limits can be scientifically measured are termed *determinate* structures, while those bridges that utilized either excessive truss members, or the early iron bridges built before materials could be tested were *indeterminate*.

As advanced as wooden bridge builders had become in their craft, the structures they designed and built were indeterminate in nature, meaning that the stress limits that they could endure could not be measured scientifically. While early iron bridge components also could not be effectively tested and iron quality varied widely, by 1870 test machinery had been invented and put into use that was amazingly accurate. One of the largest such test machines in the world, built by 1893, was located at the Phoenix Iron Company in Phoenixville, Pennsylvania, and was able to exert a stress of over two million pounds on a sample up to forty-five feet long, while a similar testing machine at a U.S. arsenal facility had the precision to measure the strength of a strand of hair and a bar of steel on a nearly equal basis (Merriman and Jacoby, *Roofs and Bridges, Part III*, pg. 83).

Top: The riveted connections on this 1924-built Warren pony truss bridge in Barre, Vermont, clearly show the rigidity of this type of bridge connection. This bridge once stood in Shoreham, Vermont, before its relocation in 2006. *Bottom:* The Union Bridge over the Pemigewasset River between Ashland and Bridgewater, New Hampshire, a Pratt deck truss bridge built in 1931, clearly shows the series of panels and the members within that constitutes its truss.

Paralleling the advances made in material testing were those made in the field of civil engineering in regards to calculating stress and load factors for a given structure. By 1870 engineers now possessed sufficient analytical means to determine measurable stresses, usually expressed in pounds per linear foot, for most standard bridge designs, and within another ten years were well able to solve more difficult problems. Such stress calculations, pioneered in part by New Hampshire's own Col. Stephen H. Long back in the 1830s, were detailed in drawings known as stress sheets and were produced in addition to original blue print drawings. These stress sheets soon became an industry standard and were utilized to measure the varying *load limits* of a bridge before its construction took place.

As to the loads carried by metal bridges, or any bridge for that matter, these are divided into two different categories, the *dead* load and the *live* load. The dead load for any bridge, simply put, is the actual weight of the bridge itself, including all its components from truss members, bracing, and flooring systems. While a bridge has to be able to carry its intended load, whether it is train, pedestrian, auto, or horse-drawn traffic, the bridge has to also be able to support its own weight. While this may seem like just common sense today, for early wooden and metal bridge builders the concept was not always understood and some bridges were overbuilt, collapsing under the stress of their own weight. Interestingly, of the many metal truss designs invented in the late 19th century, a number were found to be unsuitable because they were *too* complicated, either employing so many truss members that the bridge, at worst, was rendered an indeterminate structure, or was so complicated to build that it was considered economically unviable. Fortunately, few of these bridges truss designs were ever built.

In contrast, the live load of a bridge consists of the moving load, whether people or vehicles, passing over it. In almost all cases, the live load results in greater stresses than the dead load of the bridge due to associated vibrations generated as a load passes over. Of course, the calculations for live loads on highway bridges changed greatly in a relatively short period of time, progressing from that of largely pedestrian traffic to electric car to motorized transport traffic in the period from 1860 to 1900. Interestingly, the calculations in the last decade of the 19th century for the weights of a crowd of people ranged from 134 to 157 pounds per linear foot, surely a reflection of a leaner and better fit American populace (*ibid.*, pg. 89). What such a calculation for pedestrian weights might be today is, perhaps, a matter best left for professional engineers to contemplate!

In addition to the live loads already discussed, several other stresses that also fall into this category include those due to wind loads, especially on through truss bridges, snow load stresses, an important calculation for any northern New England bridge, and, with the advent of heavy motor traffic, impact stresses due to vibration and speed factors. With the increasing ability of civil engineers to calculate all the stresses resulting from live and dead load factors, combined with the ability to test the strength of iron and steel components, safe and dependable bridges became the norm and catastrophic bridge failures due to faulty design less common and, in the case of New England bridges, virtually unknown. Of course, the key factor in all these calculations were the design characteristics in each of the currently accepted truss forms in common use at the time of building.

There are three general types of truss bridges; the *deck* type bridge is one in which the bridge floor or roadway is attached to the upper chord and the bridge truss itself is not visible to those passing over. The opposite of this is the *through* type truss, in which the floor or roadway is attached to the lower chord of the bridge truss, with those using the crossing passing through (hence its name) the trusses. The entry (or exit) point of a bridge at either end is termed a *portal*: due to their height, through truss bridges have overhead lateral, portal, and sway *bracing* for added support. One variant of the through truss is the *half- through*

This 1936-built Warren pony truss on Dock Road in Alna, Maine, is typical of many pony truss bridges built in northern New England in the 1930s.

truss, utilized on more modern steel arch bridges such as those crossing the Connecticut River. Bridges of this type, such as the Samuel Morey Memorial Bridge between Orford, New Hampshire, and Fairlee, Vermont, are impressive in size and also require overhead bracing. In contrast, many short crossings in New England required lesser sized trusses. The *pony* or *low* truss bridge is distinguished by its utilization of smaller trusses requiring no overhead bracing and was once so commonly found in New England that their use was largely taken for granted. In more modern times the removal of bridges of this less complicated design was seldom questioned and what was once the most common bridge truss has now become an endangered form of late.

The floor of a bridge consists of *floor beams*, made of wood in early days, later iron and steel. These are connected to the bridge trusses at right angles, while *stringers* run parallel to the trusses and are supported by the floor beams. On top of the stringers is placed the road surface of the bridge; wooden planks were used for bridge road surfaces for many years and still are on some bridges. One interesting example of such a survivor is the heavily traveled Oak Street Bridge in Rollinsford, New Hampshire. However, later bridges utilized solid metal plates or, on bridges that regularly carried heavy loads, solid steel floors or reinforced concrete. On those bridges that had pedestrian sidewalks, these were cantilevered off the side of the bridge and generally attached to the floor beams and end posts with a framework of iron or steel.

As with all bridges, the stress load is ultimately transmitted to the *abutments*. These are located at each end of the structure, either on a river or stream bank, or on a built-up embankment for those bridges crossing another road or railway, and are ultimately what holds the bridge up and in place. Ideally, the face of the abutment is at a right angle to the line of bridge; however, there are some situations, especially with railroad bridges, where this is not possible. When this occurs, the bridge is on a *skew*. A *skew* bridge is easily recognizable when looking at it from its portal, where it is noticed that, in order to keep the trusses symmetrical (so that its floor beams are at right angles to the trusses), the first panel of one truss is positioned so that it is directly opposite the second or third panel (depending on the amount of skew) of the other truss. Several interesting examples of skewed railroad bridges can be found in New Hampshire, including that across the Merrimac River in Manchester, and the Broadway Street overpass bridge in Dover.

The standard method of attaching a bridge to its abutments required that one end be anchored on a metal base plate in a fixed position, while the opposite end on all but the shortest structures was mounted on *friction rollers*, allowing the bridge to move slightly to insure a uniform distribution of weight when carrying its loads, as well as allowing for seasonal temperature variations that causes metals to expand and contract. These rollers are enclosed in a protective box or "nest" so as to keep dust and dirt out and were filled with lubricating oil. On smaller trusses, the moveable end of the bridge may have consisted of a smooth iron sliding plate or a rocker plate.

The friction rollers of the pedestrian bridge in Swanton, Vermont.

The abutments to which the bridge was anchored were made from a variety of masonry materials that included simple field stones in rural settings, or blocks of cut stone, often granite, in the beginning of the iron bridge-building era, and later reinforced concrete. Where longer bridges were required, *piers* were often built up from the river bed at a central point to provide additional support. As with abutments, piers were constructed from cut stone blocks and later concrete and were constructed in a variety of configurations. By the 1890s, more cost efficient steel cylindrical piers were sometimes utilized by bridge builders as an alternative to masonry. Once common, this type of pier is now rare. One example are those in the now partially dismantled Meadow Bridge in Shelburne, New Hampshire. The longer the bridge, the more piers that might be utilized. Each segment of a bridge, whether built from abutment to abutment (without piers for support), from abutment to pier, or from pier to pier, is termed a *span*. Thus, a bridge built without a pier is termed a single span structure, that built with a single pier is a two-span structure, and so on. The Winnisquam Bridge that once connected Franklin and Tilton, New Hampshire, is an interesting example of a multi-span bridge. It is worth noting that many of the early iron bridges in New England utilized the old abutments and piers of the covered bridges that they were built to replace, a nice example of good, old-fashioned Yankee frugality!

Finally, there is the bridge *approach* itself — that portion of a road or crossing that leads to the bridge. In many situations, there is no bridge approach proper, just the roadway itself. In other situations, approaches have to be constructed so that the roadway and bridge can meet at an appropriate level. In unusual cases, as with the Meadows Bridge in Shelburne, smaller truss bridges of a type different from that of the main spans were sometimes utilized on an approach to the primary crossing, thus creating what is referred to today as a *combination truss* bridge. One of the most unusual approaches found in New England is that for the Sewall's Falls Bridge in Concord, New Hampshire. A later addition to the original bridge, the western approach consists of eight segments of steel I-beam supported ramps that gradually rise to the level of the its entrance. The River Street Bridge in Bethel, Vermont, is another structure that has a similar approach solution with five approach spans in all utilized at both ends of the bridge.

In addition to the structural details discussed above, there are also several aspects of general bridge design that merit attention, including paint schemes and nonessential decorative features. Such features as these are the most visible even to the most casual of bridge observers and ofttimes imparted to a given bridge a unique appearance that defined its image in the community it served.

The paint scheme of any bridge is extremely important and serves a dual purpose; while the colors chosen for iron and steel bridges in New England seem to have been somewhat varied, possibly due to local tastes and traditions, the very painting of the bridge itself was vital in protecting it from exposure to the elements. It is likely that nearly all of New England's early metal bridges had an undercoating of red boiled linseed oil or lead paint, but Maine seems to be unique in the region for its outwardly painted red bridges in New Sharon, nicknamed the Red Bridge (no longer red and now endangered), and the Ryefield Bridge in Harrison, a striking beauty.

The undercoating was, and still is, an industry standard among bridge fabricators (though the composition has changed), the paint being applied at the machine shop where the bridge components were manufactured. This paint was usually applied by hand in the early years for the most thorough and economical results, though paint spraying applications had been invented by the late 1800s. The final, outward coat of paint was then applied once the bridge was erected in its intended location. Often times the final painting was completed by the bridge contractor that erected the bridge, but sometimes this was done by a local firm or individuals. Afterwards, at intervals that might range from 5 to 10 years, towns and cities were faced with the necessity of repainting their bridges, primarily for preservation purposes, but also with overall appearance in mind. Indeed, the painting of bridges by local highway departments, as well as by the state, could become a seemingly constant chore, imposing a heavy financial burden on even some of the state's most prominent cities. One interesting case in point is that of the city of Dover, New Hampshire. By 1919 the city had six iron bridges, those at Lower Washington Street, Watson Bridge at Waldron's Falls, Whitcher's Falls, the Central Avenue Bridge, the Washington Street Bridge, and Sawyer's Bridge. For several years from 1919 to 1921 the city highway commissioner stressed the fact that all needed to be scraped and painted, though the funds for such were not forthcoming. The Watson Bridge had not been painted since 1912, while Sawyer's Bridge had been last attended to in 1915, the paint being purchased from the Detroit Graphite Company at a cost of $202.93. As a result of this call, the city's heavily traveled Central Avenue Bridge, a deck bridge, was coated with lead paint by a Boston contractor, M.M. Devine, at a cost of $1,798. However, the painting of the other bridges seems to have been delayed. The Watson Bridge does not appear to have been sandblasted and painted until 1931, at a cost of $303.81, though it is unclear if a new bridge was built in the interim. Towns and cities throughout the state, then as now, were often faced with tough budgetary considerations when it came to maintaining their local infrastructure.

As to the outward color of New England's metal bridges, there has been but little historical research in this area. The fact that so many of the early bridges were torn down and replaced with little or no historical surveys done, and almost no paint chip analyses, means that most of the information in this area has been forever lost. While town records frequently mention the painting of their bridges, as in Dover, none have been found that mention specific colors chosen. What little surviving evidence there is, however, suggests that at least several colors were used. A number of surviving bridges in New Hampshire, including Patterson Hill Bridge in Henniker, Thunder Bridge in Chichester, Dennis Bridge connecting Hancock and Greenfield, and Pingree Bridge in Salisbury are painted with an aluminum paint that gives them (in the case of the newly refurbished Chichester and Henniker bridges), or once did, a bright silver finish that is most attractive. The Tilton Island Bridge in Tilton, New Hampshire, also once had a silver finish, though it is now painted black. In Vermont, bridges once painted

This 1915-built Pratt truss bridge, the Patterson Hill Bridge in Henniker, New Hampshire, was restored with the silver finish typical of spans built by the Groton Bridge Company.

this color include the Mill Street Bridge in Woodstock, built by the Groton Bridge Company in 1900, while the large Pennsylvania truss structure in Swanton, Vermont, built in 1902 and relocated from its original location in nearby West Milton in 2009 has a silver finish. This aluminum paint was one of the standards in the bridge industry for many years throughout the U.S. Interestingly, many of the bridges that bear this color were built by the Groton Bridge Company of Groton, New York, making it possible that aluminum paint was a standard in their design specifications.

Another color of paint that may have been widely used is that of black. Some early photo and postcard views would seem to bear this out, especially for early wrought iron bridges and chain bridges. Among the few historic metal truss bridges to maintain this color today is the recently renovated Paddock Road Bridge in Springfield, Vermont, and the Truesdell truss supported Tilton Island Bridge in Tilton, New Hampshire. It is highly likely that many of the early cast and wrought iron bowstring truss bridges, such as that at Monadnock Mills #6 in Claremont, also bore this color scheme. One very interesting example of a bridge known to have been painted black in color was the imposing Memorial Bridge between Portsmouth, New Hampshire, and Kittery, Maine. Completed in 1923 and designed by noted engineer J.A.L. Waddell, this historic vertical lift bridge was painted with a black graphite paint that was favored by its designer. The Memorial Bridge would retain this paint scheme, which has been verified by paint chip analysis, until after World War II, when it was repainted with a standard New Hampshire Department of Transportation green color scheme. While this black paint scheme may have been unusual, it certainly proved useful during the blackout conditions imposed on American coastal cities during World War II to protect them from possible enemy attack! In more recent times there was some brief discussion about reviving this original paint scheme during future renovations, but the problem of seagull droppings marring such a starkly painted structure, and one prominent on the local skyline, made this choice impractical.

Probably the most widely used bridge color paint scheme in northern New England was that of a light green. This is due to the simple fact this color also became a national standard among bridge engineers. Recent research in New Hampshire helps to shed some light on the "greening" of that state's bridges. In May 1935 the New Hampshire State Highway Department (the predecessor to today's Department of Transportation) adopted the color for use in all structural (i.e., bridge trusses and girders) applications. However, the exact shade of green that was used by the department has been the subject of some debate, especially in regards to the rehabilitation of the Samuel Morey Bridge, a steel arch structure, spanning the Connecticut River between Orford, New Hampshire, and Fairlee, Vermont. When that bridge was built in 1936, the original specifications called for a "sage green" color; but, according to Jerry Zoller of the New Hampshire Bureau of Bridge Design, "there is no objective way to know what color "sage green" is" (letter of March 15, 2001, to Robert Landry). Interestingly, the first bridge color specified at the federal level, in 1950, was the same color that was adopted by the state in 1954 and was once again defined as "sage green." Maine and Vermont also built many bridges that utilized this same color. This light green color was thus used on regional bridges built or maintained by state highway and transportation departments for well over fifty years. However, one problem with this color is that fact that it fades or bleaches due to weathering earlier than a darker color and generally requires more frequent, and thus costly, repainting. Because of this, by 1999 New Hampshire adopted a new and darker shade of green, known, appropriately enough, as "Dartmouth green" for all bridge structures. In addition to the green color used by the state of New Hampshire, bridges at the local level throughout the region were also known to have been originally painted green, including the now gone Osgood Bridge in Campton, New Hampshire, though the exact shade used is unknown.

In concluding the discussion on bridge paint color schemes, it is important to remember that though only several specific colors have been mentioned, other colors were certainly utilized in early bridges, some of which are undocumented. Interestingly, "Federal Yellow" was another paint color adopted under national standards developed in the 1930s, but no bridges of this color have been documented in northern New England. The color grey is another that has been used elsewhere in the U.S., but not extensively in northern New England. The only example of a bridge painted this color that I have found was the Lake Champlain Bridge between Addison, Vermont, and Crown Point, New York, which was its original color when built in 1929. The color white may also have been used on some bridges, perhaps an extension of the old custom of whitewashed covered bridges that are not uncommon in the region. One intriguing possibility is a bridge in Fairfield, Maine, located near the wool mills. One postcard view shows this lenticular bridge painted a gleaming white color, but it is unknown if this was its actual color, or whether the postcard company exercised a bit of artistic license. However, it is not unreasonable to conjecture that, especially in the case of pony truss bridges that could more easily be painted, additional colors may have been used before federal standards were adopted by the states in the 1930s and sage green became the predominant bridge color across America.

Metal truss bridges have often been described, in a variety of disparaging terms, as unappealing to the artistic eye and severely lacking in ascetics. While it may be true that the beauty in most of these bridges, whether of the highway or railroad type, lies in the very geometric nature of their truss forms, some bridges were constructed with features that enhanced their appearance beyond that of a strictly utilitarian structure. Indeed, while metal bridges built after 1920 generally lack specific ornamentation beyond that of a builder's nameplate, early iron bridges in particular often included a number of flourishes. These included finials, urns, or orbs atop endposts, as well as decorative panels or cresting above portal bracing that rivaled in form intricate wrought iron fences used in public parks, upscale residential areas, and

Top: The gleaming white paint job on this two-span lenticular truss bridge in Fairfield, Maine, if accurate, is unusual compared to most bridges in the region during this time period. *Bottom*: The Whipple truss Saco River Bridge in Bartlett, New Hampshire, was interesting, if this postcard view is accurate, for its decorative overhead panel that seems to have featured a bear and other animals. This view is also interesting as it likely shows a circus wagon passing over the bridge; though barely discernible, the figure at right is dressed in a bear costume!

cemetery settings. Since few metal truss bridges constructed before 1900 remain, instances of this type of ornamentation are now relatively rare. Spectacular surviving examples include the lenticular truss bridge at Highgate Falls, Vermont, built by the Berlin Iron Bridge Company; Meadows Bridge in Shelburne, New Hampshire, constructed by the Groton Bridge Company; and Thunder Bridge in Chichester, New Hampshire, another Berlin Iron Bridge Company production. One intriguing example of ornamentation on a now lost bridge may

The Anatomy of a Bridge

Top: Thunder Bridge, Chichester, New Hampshire. *Bottom:* Highgate Falls Bridge, Highgate, Vermont.

be seen in a postcard view of the Saco River Bridge in Bartlett, New Hampshire. Though exact detail is lacking, the decorative panel above the portal bracing appears to depict animals, including what appears to be a bear, in a forest scene, which would be entirely appropriate for a bridge located in a wild and mountainous setting. Other elements found on New England metal truss bridges that were both attractive and useful include decorative lamp posts and wrought iron railings, and panels that protected pedestrians from falling into the river below. One beautiful surviving example of a crossing with such flourishes is the Elm Street Bridge in Woodstock, Vermont, a Parker Patent truss bridge.

One final decorative element that may be discussed is that of bridge builder name plates or plaques. These plates, most often mounted in prominent locations at eye level on truss beams at one end of a bridge or centered above on the portal bracing, came in a wide variety of styles and shapes ranging from the simple to those that were quite ornate. As with other design elements, their purpose was multidimensional, serving not only as decoration and a historical record of the bridge's building date, a form of advertisement for the bridge builder, as well as, perhaps, a statement of pride, of sorts, for a community that had the re-sources to cause such a structure to be built. Indeed, nowhere is that pride at having such a modern structure erected more evident than in these name plates, such as those found on the Highgate Falls, Vermont, bridge, Patterson Hill Bridge in Henniker, New Hampshire, and Ryefield Bridge in Harrison, Maine. Such plates list not only the name of the bridge builder, but also the names of the selectmen in office at the time it was constructed. In fact, name plates of this type may serve to highlight the sociological differences between metal truss bridges in general and their covered bridge predecessors.

Sign on Scribner Road Bridge, Fremont, New Hampshire. No longer extant. Signs like these were once common on early metal bridges but fell out of favor by the 1920s (photograph courtesy of Matthew Thomas and the Fremont Historical Society).

Covered bridges are what may be termed *vernacular* structures, those built on local traditions without formal plans, usually by someone within the community or with close ties. Barns are another example of a vernacular structure. In the case of covered bridges, their building was so commonplace and their builders usually well enough known, because they lived locally, such that name plates are virtually unknown. In contrast, metal bridges may be best described as *academic* structures, those that were built by a trained professional, in this case a civil engineer, and are reflective of contemporary popular design theories. Detailed builder plates with the names of local selectmen thereon engraved served to put a local stamp on a structure that was otherwise designed and largely built by those outside of the local community. Interestingly, it was not just larger metal bridges that boasted such name plates; Pingree Bridge in Salisbury, New Hampshire, is a small and otherwise unimposing structure, but its dual name plates are surely indicative of the importance of the bridge to local residents when it was built in 1893.

4

Notable Bridge Types of Northern New England

No matter which bridge type was utilized by a builder, their real importance lay in the scientifically developed design principals upon which most of them were based. A wide variety of iron and steel bridge types were built in northern New England between the years 1860 and 1940. Individual bridge types are discussed below, presented in the chronological order in which they are known to have generally appeared in the region and their unique characteristics detailed. However, it is important to remember that there was often significant overlap in the use of various bridge types and that, with the exception of the short-lived lenticular truss, most of them had a long life. In regards to truss designs, nearly all of those described below were tried and true designs. Each of them was found suitable for a particular location by an engineering professional, though the Warren and Pratt trusses by far saw the most use and were built in New England as late as the 1940s. Equally important to note are the several types of specialized non-truss bridges, such as the suspension, vertical lift, and several types of steel arch bridges, that are found in northern New England. Though bridges of these designs are not as common in the region, they are nonetheless important examples of how the use of iron and steel in bridge building evolved over the years.

The King-Post Truss

As mentioned previously, this truss is one of the oldest bridge trusses known to man. Timber structures utilizing this truss were once common in New England, but those constructed out of iron and steel seem to have been more of a rarity. This is almost certainly due to the fact that the bridges that utilized this truss were smaller in size and timber from which to construct them was readily available. However, because of their simplicity, it is also likely that metal king-post truss bridges never gained the publicity that larger and more exotic bridge types garnered. Only several remaining metal bridges of this type have been documented in northern New England, none of them on public roadways. While none are known to exist in New Hampshire, Vermont has had several examples. Bridge historian Robert McCullough has noted that at least one, made out of discarded railroad rails, was found on a private farm lane just off old Route 7 in the vicinity of Sunderland, Arlington, and Shaftsbury. It is unclear whether this bridge still stands today. Another bridge of this type, now long gone, was photographed in South Wallingford in 1927.

The state of Maine has one known iron king-post truss bridge still in existence today. Last known as the Tannery Bridge, though originally called Lothrop Bridge, this structure, originally built in St. Albans ca. 1894 to 1895 over Indian Stream, is made of wrought iron and has nut and bolt field (as opposed to factory) connections. Measuring just 28 feet long, the bridge was purchased for $377.80, possibly from a bridge manufacturer catalog, and its abutments were constructed at a cost of $425. Though the bridge was reinforced with steel stringers in the 1950s, it was in pretty sad shape by 1991 when it was removed. The bridge was subsequently renovated in 1992 and stayed in storage until the trusses (with no flooring system) were re-erected on dry land at Pittsfield, in front of the local Maine Department of Transportation facility in 1999.

The Bowstring Arch Truss

Invented by that pioneer bridge engineer Squire Whipple in 1841 and first erected in central New York at the same time, this truss type was almost certainly the first type of metal bridge widely erected in the region probably by the late 1850s. As the lone standing survivor illustrates, located in Claremont, New Hampshire, at the old Monadnock Mills complex, bridges of this type are generally characterized by a polygonal (curved) upper chord arching from the end of the lower chord at the abutments. The panels between the cords, typically numbering from four to six in number depending on the length of the bridge, each contain two vertical and two diagonal web members, and the whole structure built of wrought iron. Though variations of the web members are known to have been built elsewhere, these have not been documented in the region. Single span bridges of this type of truss, up to 150 feet in length, were once common and were very popular and cost effective for crossing small streams. The town of Claremont, in addition to its surviving bowstring truss, had at least one other type of this bridge in use over the Sugar River, while nearby Newport, New Hampshire, had at least one example, built after 1870, that survived into the 1980s. Other bridges of this type known to have been built prior to 1870 include three in Bennington, Vermont, two in Maine at Biddeford and Portland, and those in New Hampshire at Exeter, Manchester, and Wilton. All were built by the Moseley Iron Building Works of Boston, and many were probably Thomas Moseley's version of the bowstring arch, patented in 1857. A number of other arch bridges were also built in the region by the Wrought Iron Bridge Company of Canton, Ohio.

While New England's one surviving standing example of this type is found in a mill setting, those still extant in other states are generally located on country roads that see minimal traffic. Another interesting survivor of sorts of this type of bridge in the region is the former Murphy Road Bridge in Bennington, Vermont. This bridge stood in its original location from ca. 1860 until 1958 when it was removed by the city. Originally destined for the Smithsonian Institute in Washington, D.C., as part of a technology display in development, the bridge was later deemed too large for display and its truss components instead ended up in storage at the Bennington Museum. Though still in good condition, the old bridge is still waiting for a new home over fifty years later!

The Pratt Truss

One architectural historian refers to this truss' designer, Thomas Pratt, as "the most thoroughly educated American bridge builder at the beginning of the railroad age" (Condit, pg. 109), and rightly so. He patented a modified version of Amasa Stone's version of the Howe truss in 1842, and two years later, along with his father Caleb, patented a similar truss with a polygonal top chord, or parallel chords. Intended for either wood or iron construction, the earlier forms of this truss resemble a series of boxed Xs, while later versions of the truss utilized single diagonal members in all panels except those in the center. Significantly, the first iron Pratt trusses were built by the Pennsylvania Railroad beginning in 1850, though they were first built using wood. Even the first iron Pratt truss bridges were strengthened by arch ribs, a practice that ended by 1870, by which time the Pratt truss was proved on its own merit to be a tried and true form.

The Tannery Bridge once stood in St. Albans, Maine, but was later removed and now graces the grounds of a Maine DOT facility in Pittsfield, Maine (photograph by R. A. Wentzel, courtesy of Maine Historic Preservation Commission).

The oldest known Pratt truss bridges in northern New England still standing are that across the Connecticut River between Maidstone, Vermont, and Stratford, New Hampshire, and the abandoned Thompson's Crossing Road Bridge between Antrim and Bennington, New Hampshire, both built in 1893. However, the older Howard Hill Road Bridge, built in 1890 in Cavendish, Vermont, is now in storage and awaiting a home as part of Vermont's Historic Bridge Program. Bridges of the Pratt form were popular among New England builders for many years, remaining a common form well into the 1930s.

The Whipple Truss

This trapezoidal truss form was yet another invention by that pioneer bridge engineer and designer, Squire Whipple. This truss was patented by Whipple in 1847, the same year that his groundbreaking study, *A Work on Bridge Building*, was published, and soon gained widespread acceptance, especially with railroad bridge builders. Indeed, from 1850 to 1870 it was the most popular bridge form before being eclipsed by the Warren truss and variants of the Pratt truss forms, though some bridges of this type appear to have been built into the 1890s. Now a rarity nationwide, with none left in northern New England, specific examples of Whipple truss bridges in Vermont are unknown, while three examples — a Grand Trunk Railroad bridge in South Paris, a highway bridge in Bowdoinham, and one built by the King Iron Bridge Company of Cleveland, Ohio, between Brunswick and Topsham — have been identified in Maine. Examples of this bridge form that once existed in New Hampshire include the Eastern Railroad's Saco River Bridge and the Saco River Bridge for road travel, both in Bartlett, and possibly the Lottery Bridge in West Claremont. Though further examples of Whipple truss bridges have not been identified in the region, numerous other examples were undoubtedly built, probably the vast majority by railroads.

It is interesting to note that while the Whipple truss fell out of favor by 1890, recycled examples of this bridge likely continued to serve on secondary roads well into the 20th century throughout the northeast. The cast iron Whipple truss bridge in Bowdoinham, Maine, crossed the Cathance River and was in service there from 1895 to ca. 1950 before it was torn down and sold for scrap. This 144-foot long bridge, built by John Hutchinson of Troy, New York, in 1866, had originally stood somewhere in New York, Massachusetts or Connecticut before it was taken down after serving for nearly 30 years, no doubt replaced by a more modern and substantial structure, and subsequently sold to the town of Bowdoinham by the Berlin Iron Bridge Company. It was thus re-erected at its new location by "engineer-salesman" John Towne at a cost of $3,886 and went on to serve for another fifty years. It is significant to note that bridge engineer Llewellyn Edwards wrote of this bridge in 1934 that it "is the most interesting metal bridge structure now existing in Maine" (Connors, pg. 2).

The Warren Truss

One of the most innovative and long-lived of the iron truss bridge designs to gain popularity in America was patented by two English engineers, James Warren and Willoughby Monzani, in 1848. Universally known as the Warren truss, this simple design originally employed no vertical members or posts and resembled a series of equilateral triangles. Interestingly, according to historian Carl Condit, it is the only truss form found in nature, mainly the bone structure in the wings of several large flying birds. Subsequent versions of the Warren truss utilized upright posts, or a second set of diagonal members, the later form resembling a distinctive series of diamond-shaped patterns. This second set of diagonals resulted in modified Warren trusses known as the double-intersection Warren truss, while still other forms with more elaborate webbing are referred to as multiple-intersection Warren trusses. In contemporary accounts these forms were often called lattice trusses, no doubt due to their appearance and close resemblance to the Towne lattice truss of covered bridge fame. The oldest Warren truss bridge in Maine is of the double intersection variety, the Ryefield Bridge in Harrison, built in 1912.

First built in the U.S. in 1850, ironically by Squire Whipple, who appears to have had no prior knowledge of its invention in England, the Warren truss became popular by the 1860s, and by the end of the 19th century the Warren truss was one of the most trusted and widely used forms in American bridge building. Though the exact figures are unknown, hundreds of Warren truss railroad and highway bridges were built in northern New England between 1890 and 1940. The oldest extant Warren truss bridge is the recently renovated Foundry Bridge in Tunbridge, Vermont, built in 1889. Most of the older Warren truss bridges are of the multiple-intersection variety, including the now bypassed Medburyville Bridge in Wilmington, Vermont, built in 1896, and the previously mentioned Ryefield Bridge in Harrison, Maine. The 1892 built Rice Farm Road Bridge in Dummerston, Vermont, is an interesting and rare example of a Hilton truss bridge, an elaborate and patented version of the multiple-intersection Warren truss.

The Lenticular Truss

One of the most exotic of the early metal bridge trusses remaining today due to its shape, the lenticular truss is well worthy of extended discussion. Interestingly, nearly every bridge of this type built in America, from New England to Texas, was built by one company, the Corrugated Metal Company, known after 1882 as The Berlin Iron Bridge Company of East Berlin, Connecticut. The lenticular truss was first built in Europe in the 1850s, including Friedrich

Augustus von Pauli's bridge over the Isar River, built in 1857, and Isambard K. Brunel's 1859 Royal Albert Bridge at Saltash, England. Though Edwin Stanley did patent a lenticular truss in America in 1851 combining timber and wrought iron chords, no bridges of this type seem to have been actually built. Based on the bridges being built in Europe, William Douglas, a West Point graduate from Binghamton, New York, was granted a patent on a lenticular truss in 1878, which he alternately called a parabolic or elliptical truss. As originally designed, this type of truss had hipped (not curved) upper and lower chords, with single diagonal members inclined toward the center in each panel, except for the two center panels, which had counter diagonals resembling an X in each. No examples of this form remain in existence today, and it may be that relatively few were built, as compared with later examples of the Berlin Iron Bridge Company's work. One known example, identified by historian Robert McCullough and photographed by the Vermont Agency of Transportation in 1923, is a bridge at an unidentified location in Jay Corners, probably over Jay Branch, in northern Vermont.

William Douglas soon sold his patent to S. C. Wilcox of the Corrugated Metal Company, but that company's bridge division quickly changed its lenticular truss form after 1878, according to historian Thomas Boothby, with the addition of civil engineer Charles Jarvis to its staff upon his graduation from the Sheffield Scientific School at Yale University in 1877. Indeed, Boothby comments that Douglas' original patent drawings "are very unsophisticated and show discernible engineering errors" (Boothby, pg. 4), while Jarvis was a trained expert who went on to become the company's chief engineer and president. Based on his studies at Yale, Jarvis favored the use of a specialized lenticular form known as the Pauli truss, employing matching upper and lower chords which are curved in form and utilize counter diagonal bracing in all but the two end panels.

Though widely built, this lenticular form was relatively short-lived because it could not be stiffened enough, even with overhead bracing, to cope with the heavy loads generated by increasingly modern methods of transportation. It is interesting to note, however, that lenticular, or parabolic trusses, were also projected for use in railroad bridges. How widespread their use in this capacity in northern New England is unknown, but likely not extensive. One possible example was in Nashua, New Hampshire, where the Concord Railroad needed a crossing over the Nashua River at Taylor's Falls in 1884, where a wooden bridge "was condemned the past year, and an iron parabolic truss bridge will at once displace it" (New Hampshire, 40th Annual Report, pg. 25). While a lenticular truss bridge was indeed built, it only saw electric streetcar and regular highway traffic and no train use.

Due to the Berlin Iron Bridge Company's aggressive marketing in New England, the lenticular truss saw its heyday during 1877–1895, thereafter replaced by heavier built bridge forms. Vermont once had a large number of these bridges, including examples at Highgate Falls, Saxton's River, Richford, Woodstock, Bethel and Jay Corners. Today, only the Highgate Falls Bridge survives in its original location in Vermont, though the Bethel bridge is in storage and, hopefully, within the near future will be re-erected on the campus of the University of Vermont at Burlington by engineering students. The state of Maine has no surviving lenticular truss bridges today, but known examples of the past include those built at Buxton, Yarmouth, Bingham, Fairfield, Houlton, and Clinton. New Hampshire has the most examples of surviving lenticular truss bridges in northern New England, including the abandoned Livermore Falls Bridge in Campton, the Dow Avenue and Delage bridges in Franconia, and Thunder Bridge in Chichester. The Dow Avenue Bridge in Franconia is the only lenticular truss bridge in the region open to vehicular traffic today. Lenticular bridges of the past in New Hampshire include the large Granite Street Bridge in Manchester, the Depot Bridge in Antrim, the Taylor's Falls Bridge in Nashua, the Mason Street bridges in Berlin, and the Interstate Bridge over the Connecticut River between North Stratford, New Hampshire, and Bloomfield, Vermont.

The Baltimore and Pennsylvania Trusses

The influence of railroad companies in advancing the state of bridge engineering and truss designs is certainly evidenced by the Baltimore and Pennsylvania truss forms. Both originated from the drawing boards of engineers for the Pennsylvania Railroad about 1870, and though at first used for their crossings, were soon widely employed by other railroads and highway bridge builders throughout the United States. Both are sometimes also referred to as Petit trusses, based on the fact that the Baltimore truss was early on, in 1871, referred to as "Pettit's Stiffened Triangular Truss" (Merriman and Jacoby, *Roofs and Bridges*, Part I, pg. 222). Though the Baltimore truss was designed by the engineers of the Pennsylvania Railroad, its designation today is based on the fact that many were erected by the Baltimore Bridge Company. Both of these closely related trusses are a variant of the Pratt truss, with each bridge panel being further subdivided with the use of simpler half-diagonal and half-vertical members, effectively acting as secondary king-post trusses within the original Pratt truss. The Baltimore truss is trapezoidal in form, with parallel upper and lower chords, while the Pennsylvania truss is distinguished by its polygonal (curved) upper chord. Though originally built as railroad bridges, these truss forms were soon used in a wide variety of applications, including highway and factory bridges, and were built well into the 20th century in northern New England. The primary advantage of these bridge forms were their simpler design, compared to earlier railroad bridge trusses, that made them easier to erect and more economical in materials, yet at the same time they provided a more rigid and solid structure. While the earliest examples of these bridges were built of iron, steel soon came into use.

Baltimore and Pennsylvania truss bridges were widely employed by New England railroads, but highway bridges of these types in northern New England never gained widespread use, perhaps because of the region's more rural landscape. Nonetheless, some interesting examples still survive. Vermont has two Baltimore truss bridges, the recently renovated Paddock Road Bridge in Springfield and the Granite Street Bridge in Montpelier. This later bridge is a perfect example of the types of industrial conditions in which these bridges were often employed. The aptly named Granite Street Bridge directly served (and still does to a lesser extent) an area in Montpelier where the granite sheds were located and had to be strong enough to bear the truck loads of granite, both in quarried and finished states, that regularly used the crossing. Vermont also has three Pennsylvania truss bridges, including the 1900 built Mill Street Bridge in Woodstock (now strangely modified), the 1902 built structure that once stood in West Milton and was recently renovated and re-erected in Swanton, and the heavily traveled Route 2 bridge in Richmond, built in 1929.

New Hampshire's examples of these bridge forms include two abandoned Baltimore truss bridges, the 1903 built combination railroad and automobile bridge over the Connecticut River between Woodsville and Wells River, Vermont, and the ca. 1905 Sullivan Machine factory bridge in Claremont, as well as a large number of railroad bridges. Perhaps the best known of these are those located on former Maine Central and Boston and Maine lines in the White Mountain region in Ossipee, Conway, Glen, Bartlett, and Hart's Location. Most of these bridges were built in 1906 to 1907 and see limited use today by the Conway Scenic Railroad. A number of other Baltimore truss railroad bridges are also located on the Plymouth and Lincoln Railroad, including one in Campton, and two each in Thornton and Woodstock. Of the Pennsylvania truss form, New Hampshire has three remaining examples, all crossing the Connecticut River, at Hinsdale, Charlestown, and Piermont, all built from 1920 to 1930.

Maine, too, has its share of Baltimore and Pennsylvania truss bridges. The oldest known Baltimore truss bridges in the state were all built for railroad use; the St. John Street underpass in Portland, the 1928 built Elm Street Bridge in Biddeford, and the Free-Black Bridge between

Brunswick and Topsham. While the example in Portland, built in 1890, is the oldest metal truss bridge still extant in Maine, the 1909-built Free-Black Bridge is by far the most unique of all the Baltimore truss bridges in New England. Employing a single lane deck for auto travel suspended beneath the railroad bridge, it is a thrilling experience to travel this bridge. Of Pennsylvania truss bridges, Maine has more surviving examples than New Hampshire or Vermont, though its numbers will soon dwindle considerably. Of five such bridges, the 1916 built bridge in New Sharon is the oldest; though now bypassed all restoration efforts to date have failed due to a lack of funding. The other examples of this bridge form in the state, built from 1921 to 1929, are the international bridges at Fort Kent (scheduled to be replaced soon) and Madawaska, Stinchfield Bridge in Leeds, and the Piscataquis River Bridge in Howland, also scheduled for replacement as of this writing. As in industrial or heavy-load settings elsewhere in northern New England, the use of the Pennsylvania truss for bridges across the St. John River between Maine and Canada at Fort Kent and Madawaska, imposing and important structures which have seen heavy use over the years, is especially appropriate.

The Parker Truss

Like many truss designers of his time, Charles H. Parker was an up and coming mechanical engineer who designed his own truss form with very little actual bridge building experience to his credit. He was the chief design engineer for the short-lived Solid Lever Bridge Company, which was organized in Boston in 1867 and went out of business in 1871. This company built iron and wood cantilever bridges, with Parker's design utilizing wrought iron Warren trusses for the cantilever arms. In 1870 Parker patented a truss which had a curved upper chord, sloping endposts, and pin connections. Though this truss form outwardly resembled Squire Whipple's bowstring arch truss, Parker's version was stronger and more economical in its use of materials. Parker was later a designer for the National Bridge and Iron Works in Boston, and an unknown number of his Parker Patent Truss bridges were built in New England. In fact, there are just a handful in existence today in America, and only three, all in Vermont, have been documented in northern New England. One was built in Springfield in 1869, before Parker was granted his patent, but it soon fell to flood waters. The other two known Parker Patent bridges, both constructed of wrought iron, in Vermont are still standing today, though neither functions in their original form. These are the beautiful Elm Street Bridge in scenic Woodstock, and the Vine Street Bridge in Northfield. The later bridge was purchased by the Vermont Central Railroad and a publicity photo taken in 1870 shows it bearing the weight of two locomotives. Today, this significant structure, moved from its original location, serves pedestrian traffic only.

While Parker's Patent Truss was a well designed form, it does not seem to have been widely built, perhaps due to heavy competition from the salesmen of the Berlin Iron Bridge Company. However, Parker's name has also been attached to a variant of the Pratt truss that employs a polygonal upper chord with many jointed angles. Yet another variant of this truss is called a camelback truss, which utilizes a polygonal upper chord with five segments, making it easier and less costly to build. Bridges utilizing the Parker and camelback trusses were popular in northern New England and were built well into the 1930s. While many bridges of this type were built in northern New England as replacements for bridges lost during the flooding of 1927 and still serve traffic, a few older Parker truss structures are still extant, including the 1905 built Chef Road Bridge (its trusses now ornamental) in Jackson and the large 1907 Depot Street Bridge (now abandoned) over the Merrimac River in Boscawen, both in New Hampshire, and Vermont's Mill Street Bridge in Cavendish (1905) and Rabbit Hollow Bridge in Northfield (1908), both of which are pony truss bridges.

The Lake Champlain Bridge between Vermont and New York was revolutionary for its continuous truss design.

Continuous Truss Bridges

A continuous truss bridge is one whose truss (whether it be a Warren or Pratt form) extends over two or more spans, as opposed to the standard two or more *separate* trusses often employed on shorter multi-span bridges, and is thus supported in at least three points, or more depending on the length of the bridge. Though bridges of this type have been built since the 1870s, they were fairly rare until the 1920s and usually utilized on large railroad bridges. Continuous truss bridges were not widely used previously mainly due to the fact that the science of stress analysis had not progressed far enough to give builders confidence in their use. However, with advances in bridge engineering and steel production, continuous truss bridges came into greater use by the late 1920s, their use further encouraged by the rapid increase in auto transportation and their economical use of materials in large structures. Since continuous truss bridges were often employed on highway bridges over large bodies of water, their use in northern New England has been limited. Probably the best known continuous truss bridge in the region was the noted Lake Champlain Bridge. Completed in 1929, it measured 1,014 feet in three spans and was a vital link in the transportation network between Vermont and New York before its rather sudden demise in 2009.

There are only four historic (pre–1940) continuous truss bridges remaining in northern New England, three of which are still in highway use. The most impressive of these is the 1934-built General Sullivan Bridge that spans Great Bay between Newington and Dover Point in New Hampshire, but it has been bypassed since 1966. While it has been a favorite of exercise enthusiasts, fishermen, and sightseers, this bridge is badly rusted and its days seem numbered. The three continuous truss bridges still in use are all in Maine, and were built in 1937 as flood replacement bridges, the Bars Mill Bridge in Hollis, the Durham Bridge between that town and Lisbon, and the West Buxton Bridge in Buxton.

Cantilever Truss Bridges

This type of bridge is also a rarity in northern New England, partly because the cantilever truss is generally used for larger structures. The primary factors involved in choosing a cantilever truss include circumstances where falsework (the supports and scaffolding utilized during a bridge's construction), were a regular truss to be used, would be extensive and add greatly to construction costs, or where pier construction might be problematic, perhaps due to the depth of water and the grade of rock below or strong currents. A typical cantilever truss bridge has arms coming from each end of a river bank or shoreline that are connected at mid-channel by another truss suspended between them. Thus, the cantilever truss is supported at only one end of the bridge and at a point in-between, rather than at both ends like a regular truss bridge; that arm of the bridge between the supports is called the anchor arm, the other the cantilever arm. The stress distribution in a cantilever truss is exactly opposite that of a regular truss bridge, its top chord being in compression and the bottom chord in tension. Wooden cantilever bridges, albeit of simpler construction, date back hundreds of years (one example was built in India in the 1600s) in Europe and Asia, and though discussed in theory by American builders, did not begin to see extensive use until the 1870s, mainly by railroad engineers.

Four notable examples of cantilever bridges can be found in northern New England, though they were all built after World War II, three of them in Maine, and one continuous truss bridge with cantilevered anchors in Vermont. New Hampshire has no examples of this bridge type. Those in Maine are the Augusta Memorial Bridge in the state's capital, the Max L. Wilder Memorial Bridge over the Sasanoa River in Arrowsic, and the Aroostook River Bridge in Caribou. Of the three, the first, a four-span deck structure over the Kennebec River built in 1948 to 1949, is the most impressive. However, the Max L. Wilder Memorial Bridge, built in 1950, is a beautiful example of the principles of the cantilever truss. The 1952 built Aroostook River Bridge was the third and final example of this bridge type ever built in Maine, a critical and attractive link in the state's northern transportation network. Vermont's sole bridge utilizing cantilever principles is the I-91 bridge over the West River in Brattleboro, built in 1960. This continuous truss bridge has cantilever anchor arms on each end that meet at center span, resulting in what Vermont bridge historian Robert McCullough calls "a graceful arch" (McCullough, pg. 143).

Suspension Bridges

This bridge type, one of the oldest known to man, was one that was greatly influenced with the advent of iron and steel. Early on in human history, especially in Asia, suspension bridges were widely built. And though at first they were crude affairs, constructed of strong vines, and later on rope, there is evidence that in China suspension bridge builders were using iron in their construction by A.D. 580 (Condit, pg. 163). In contrast, European builders did not conceive the idea of a suspension bridge until centuries later, likely inspired by illustrations of Chinese suspension bridges. The first example built in Europe was in 1741 in England.

The first suspension bridges in America were the chain bridges built by Pennsylvanian James Finley, as previously mentioned, in the first decades of the 19th century. Finley's bridges were an important development, for he was the first to create a level-deck suspension bridge with stiffening members and a rigid deck, as opposed to early suspension bridges whose deck followed the curve of the bridge itself. These chain bridges eventually came to be replaced by suspension bridges utilizing wire-rope cable. But the first suspension bridge with wire-rope cable, built in Philadelphia, collapsed the same year it was built, 1816, causing wire-rope to be suspect for a number of years. While French engineers further pioneered the development and construction of wire-rope suspension bridges, these ideas would not be applied in America until the 1840s, pioneered by Charles Ellet, Pennsylvania-born, but educated and influenced by a year of engineering study in Paris. However, while Ellet may have begun the re-introduction of the wire-rope suspension bridge, it was German-born engineer John Roebling who not only perfected the form, but in the process became one of America's premier engineers and bridge builders. Though John Roebling, and later, his son, Washington Roebling, are best known for building the legendary Brooklyn Bridge, he actually constructed many notable bridges in New York and the Midwest, and his steel wire cable operations in Trenton, New Jersey, are still in operation today.

The suspension bridges of northern New England, while few in number and not nearly as historic or groundbreaking as the Brooklyn Bridge, are nonetheless fascinating survivors. Nearly all are examples of factory or town pedestrian bridges, except for two large suspension bridges built by noted engineer and bridge builder David B. Steinman on the Maine coast. Vermont has no historic suspension bridges remaining today, but once had several interesting larger examples of this type, as well as a number of smaller pedestrian suspension bridges that have also vanished. The largest of these was the 1887-built Bancroft Falls Bridge at Sheldon Springs, measuring 250 feet in length, which survived until the flood of 1927. A more unusual example was the Sutherland Falls Bridge in Proctor, measuring 122 feet long and built ca. 1900 by the Proctor Marble Company to carry a slurry pipeline across Otter Creek. This bridge stood until the late 1990s, when it was burned down.

New Hampshire, too, has had a few suspension bridges in its day, though today only two examples remain. Notable bridges of the past include the Chesterfield-Brattleboro Bridge between Chesterfield and Brattleboro, Vermont, across the Connecticut River, built in 1889. Measuring 320 feet long, the bridge served both communities well until it was damaged by a heavy load in late 1935, and was subsequently carried away in the flooding of 1936. Of this bridge's final construction costs, Brattleboro paid thirty-five percent of the $12,675 bill, with Chesterfield paying the remainder, an equitable division when it is remembered that New Hampshire owns the Connecticut River to the high-water mark on the Vermont side! Another notable suspension bridge was one with wooden towers, a not uncommon feature in New England suspension bridges, that stood in Gorham for many years. Today, the only surviving examples are those in Milford and Littleton. The latter is an undistinguished pedestrian bridge formerly utilized by factory workers, while the Milford bridge, built in 1887 by the Berlin Iron Bridge Company, is a beautiful example of a pedestrian suspension bridge. Because of the swaying motion one encounters when crossing this bridge, it is known locally as the "swinging bridge," a common designation for bridges of this type in the region.

While New Hampshire has its examples, it is in Maine, which has five historic suspension bridges, that we find the most interesting and widest range (big and small, old and modern) of this bridge type in northern New England. One of the oldest standing suspension bridges in the United States is the Wire Bridge in New Portland, built in 1866. Designated by the Maine Section of the American Society of Civil Engineers as an engineering landmark, the Wire Bridge is a notable and exciting example of the bridge-builder's art practiced in the Maine wilderness! Two other notable suspension bridges, both pedestrian in nature, are the Two Cent Bridge in Waterville, and the Brunswick-Topsham footbridge. The former was a toll bridge built in 1903 for the Ticonic Footbridge Company and served local factory workers; the later was built in 1892, rebuilt after the flooding of 1936, and utilized wire-rope cable produced at the Roebling Works in Trenton, New Jersey. While the previously discussed Maine suspension bridges are modest in size, two others, the Waldo-Hancock Bridge, connecting Prospect and Verona, and the Deer Isle Bridge, are northern New England's closest relatives to San Francisco's famed Golden Gate Bridge. Both these Maine bridges were built by one of America's foremost bridge designers and builders, David B. Steinman. The Waldo-Hancock Bridge he considered among the finest examples of his work, while the later, albeit considered less of a beauty, was a solid example of his work constructed at a bargain price!

Steel Arch Bridges

The arch bridge, in all its varied forms, is probably the most widely recognized bridge type by the general public, and certainly one of the most aesthetically pleasing. The previously discussed bow-string arch bridge, patented by Squire Whipple and others in the 1860s, was a fixed-arch structure. Bridges of this type transmit their stresses along the length of the arch to the abutments upon which they are anchored and are indeterminate, meaning that the stress loads they were capable of bearing could not be calculated in these early days of civil engineering (though now they can be). Despite this fact, many notable bridges of this type were built in epic proportions, including the landmark Eads Bridge in St. Louis (1874), in the U.S. for use in both highway and railroad structures. Modern steel arch bridges with more advanced engineering incorporated into their design came into use by the early 20th century, along with advances in steel production.

One of these first advances resulted in the *tied arch* structure. Though tied arches were actually first used to support the arched ceilings in large railroad buildings beginning in the 1870s, they were later applied to bridge construction. A tied arch bridge is distinguished by its floor panels passing through the arch and suspended from it by hangars or ties made of iron, first, and later steel.

Also utilized in these gigantic railroad buildings was the newly developed concept of the *hinged arch*. Like many railroad technological advances, these ideas were soon applied to other structures, eventually resulting in a number of notable hinged arch bridges being built in northern New England, and elsewhere, that combined beauty and measurable stress loads. These hinged arch bridges came in several different forms. The *single-hinged* arch, the least popular of this bridge type, incorporates a hinge in the middle point of the arch at its crown, or highest point. A *two-hinged* arch locates a hinge at each end (known as springing points) of the arch near the abutment. A *three-hinged* arch bridge has hinges at the crown and at its spring points. In each of these types of hinged arch bridges, the division of the arch meant that the forces acting against the hinge could be measured, resulting in a determinate structure. Both deck and through arch bridges were built, the through arch bridge, similar to the through truss bridge, having overhead lateral bracing to strengthen the structure. In early steel arch bridges, the arches were often reinforced by truss forms, such as the Pratt truss, while later, more advanced designs utilized hollow ribs of steel to stiffen the structure, resulting in a box-girder-rib arch bridge. Hinged bridges of epic proportion were first built in northern New England in 1905, with the construction of the Walpole-Bellows Falls Bridge across the Connecticut River between New Hampshire and Vermont. Costing just $47,000 and built by notable engineer Joseph Worcester, this three-hinged arch bridge was the longest in America at the time, spanning 540 feet. Later arch bridges were built across the Connecticut River and elsewhere in the region, spurred on by two major factors, the primary one being the need for sturdy flood replacement bridges. However, advances in American transportation also provided encouragement for the building of these beautiful arch bridges; because millions of Americans were taking to the roads as automobile ownership increased exponentially, not only were better and stronger bridges needed, but also more attractive ones! It is no mere coincidence that many of the Connecticut River arch bridges were soon designated as "most beautiful" bridges in their class upon completion and quickly became regional landmarks for local residents and tourists alike.

Approximately ten steel arch bridges are known to have been built in northern New England, one in Vermont, two in Maine, five across the Connecticut River between New Hampshire and Vermont, and two others in New Hampshire. Vermont's sole example of this type of bridge still standing is the Quechee Gorge Bridge over the Ottauquechee River in

Top: The Arch Bridge over the Connecticut River between Walpole, New Hampshire, and Bellows Falls, Vermont, was a wonder in its day. Designed by J. R. Worcester, it was the largest bridge of its kind in the world upon completion. *Bottom:* The soaring Notre Dame Bridge across the Merrimack River in Manchester, was also noted for its concrete arch approach spans and Art Deco abutments (seen at left) (courtesy of New Hampshire DOT).

Hartford. This deck arch structure, designed by John and Edward Storrs, was built in 1911 and originally served as a railroad bridge until being converted to highway use in 1933. Maine's sole surviving steel arch bridge is the Morse Street Bridge in Rumford, built in 1935, though a similar structure once stood in the town of Turner as well. All of New Hampshire's remaining steel arch bridges, four in number, cross the Connecticut River and were built from 1923 to 1938. These will be discussed at further length in a later chapter. Joseph Worcester's first steel arch bridge across the Connecticut stood from 1905 until 1982; famed when it was built, it gained some fame at its death as well. Closed since 1971, the bridge was scheduled for demolition in December 1982, but still stood after five demolition charges failed to bring down the structure, gaining in the process a fair amount of national publicity and raising questions as to why the bridge had to be demolished when it was strong enough to withstand demolition expert's best efforts. Eventually, however, key bridge members were cut and the old bridge, a fighter to the last, fell into the river, gone forever. New Hampshire had two other arch bridges solely within her borders that no longer remain. One was a steel tied arch bridge across the Pemigewasset River in Woodstock, the other the famed Notre Dame Bridge across the Merrimack River in Manchester. This later bridge was a two-hinged arch built in 1937 at a cost of $732,000, measuring in at 444 feet long. This notable structure was a distinguishing feature on the Queen City skyline for years until it was demolished in 1989. Like the Walpole-Bellows Falls Bridge demolished seven years earlier, the loss of this bridge was later lamented as a short-sighted decision that deprived the city of one of its most recognized landmarks.

Moveable Bridges

Bridges of this type, like so many other bridge forms, date back to antiquity. As their name suggests, moveable bridges are, quite simply, those bridge structures whose roadway can be moved in order to allow the passage of water borne traffic. In olden days in Europe, bridges of this type, such as a drawbridge across a castle moat, were designed for defensive purposes. However, in America moveable bridges have been built over waterways since colonial days to accommodate traffic by land and by sea. Originally operated by hand, such bridges were, and still are today, an important compromise between the needs of the maritime trade and that of overland trade and everyday transportation. Of the three types of moveable bridges, the *bascule bridge*, the *swing bridge*, and the *vertical-lift bridge*, all were built in northern New England, albeit in relatively small numbers except in coastal areas. It is important to note that all of these moveable bridge types utilized a number of truss forms within their construction. These bridges are also quite interesting in that, when operating as designed, they require manpower, in the form of bridge operators, to keep them moving. Thus it is, that though the days of the old-fashioned lighthouse keeper of New England are long gone now that lighthouses are fully automated, we still have men and women in coastal areas that carry on similar duties, albeit in less isolated fashion, of keeping those at sea, and those on shore as well, safe and on the move day and night, no matter what the weather conditions.

The bascule bridge is one of the oldest types of the moveable bridge, and one closely related to the drawbridge of old. Early versions consisted of a single span hinged at one end and moved by a rope or cable attached to the free end that ran upward and shoreward to its source of lifting power. In premodern times this power source was often a hand-cranked windlass. Early bascule bridges were limited in use because they did not have counterweighted balances to aid in their operation. However, modern bascule bridges utilize counterweights and power driven machinery located at the fixed point of the span, eliminating the need for ropes and cables. Modern large scale bascule bridges evolved into two major types (though many other types were also patented), these being the simple trunnion or "Chicago" type, and the roller-lift type, both found in the region, the later being distinguished by their tower structure. Bascule bridges in northern New England were generally of the single "leaf" span type, with one span being moveable, due to their smaller applications, though double-leaf span bascule bridges were also built on larger waterways.

Historic bascule bridges in this region are rare today, and but two are active. One of these is the steel bascule bridge, a double-leaf structure over an 80-foot channel built in 1953 on Lake Champlain between Grand Isle and North Hero in Vermont. Another bascule bridge, built in 1938 over Missisquoi Bay between Swanton and Alburg, notable as the state's first highway bascule bridge, was replaced several years ago.

Maine has one historic highway bascule bridge remaining, the Wadsworth Street Bridge in Thomaston. Built in 1928, this bridge was of the roller-lift type and is distinguished by its curved Pennsylvania truss tower and Pratt truss through span. Sadly, this bridge was fixed in place in 1966 and all its machinery removed. However, its rarity makes it an easy candidate for continued preservation. Interestingly, while several other bridges of this type were earlier built in Maine, the building of the Thomaston bridge is something of a mystery, as the roller-lift type of bascule was considered obsolete by the 1920s due to its complicated design and maintenance difficulties. It was probably for this very reason that the bridge seldom opened after 1950. Of Maine's other bascule bridges, one is a 1922-built bridge on the Eastern Railroad spanning St. George River, while the other is a modern structure, the large Fore River Bridge in Portland, completed in 1998.

New Hampshire has two bascule bridges remaining, these being the Little Harbor Bridge between New Castle and Rye, spanning 43 feet over a back channel of the Piscataqua River near the famed Wentworth-by-the-Sea Hotel, built in 1942, and the Neil R. Underwood Memorial Bridge (formerly the Hampton River Toll Bridge) in Hampton, built in 1949. The former bridge, if it still works, has not been opened for many years. However, the Hampton River bascule, which operates from April to October, may make as many as 20 lifts a day during the height of the tourist season. After much debate, this bridge underwent extensive renovations in 2010, though many locals desired a fixed replacement bridge built higher over the water to alleviate traffic congestion. Another bascule bridge in New Hampshire was the Alexander Scammell Bridge, built in 1935, connecting Dover Point and Cedar Point in Durham where Little Bay meets Back River. This bridge was seldom opened in its last decades and was replaced with a fixed bridge ca. 2002. According to State Architectural Historian James Garvin, after the loss of the Scammell Bridge, the New Hampshire DOT pledged to preserve the state's remaining two bascule bridges, a promise they kept in regard to the bascule bridge in Hampton.

A second type of moveable bridge is the swing bridge. Like the bascule this bridge type also has a long history in America, with early examples being built of wood. The most common type of this bridge in its mature form is one that pivots on a center bearing and moves, or swings, a span crossing over a navigation channel to a point where it rests at a 90 degree angle to the fixed portion of the bridge thereby opening the shipping channel to traffic. Swing bridges have a few disadvantages when compared to the bascule type; including the fact that they require a center pier that can constrict a waterway and pose a potential hazard to shipping, the greater length of time it takes for them to operate in opening a waterway, and the fact that they have to be fully deployed to allow shipping to pass, while a bascule can be partially opened if needed, depending on the height of the vessel transiting the waterway. Despite these factors, the swing bridge had one huge advantage that caused them to be built in larger numbers in northern New England than the bascule, and that was their initial construction costs and generally lower maintenance costs over the life of the bridge.

The number of swing bridges constructed over the years in the region is unknown, but probably runs to several dozen. All of the known swing bridges built in Vermont, eight in number, were located on Lake Champlain; half of these were built for the Rutland-Canadian Railroad's Island Line, while the other four were highway bridges. None of these bridges have survived though the last example, the Alburg-Rouses Point Bridge, stood into the 1980s until it was replaced with a fixed bridge. New Hampshire, too, has no surviving swing bridges, though at least three are known to have been built in the state. One was located in the Lakes Region near the outlet to Winnisquam Lake in Laconia. However, the best-known of these was the Newington-Dover Point Bridge over the waters connecting the Piscataqua River with Little Bay. This unusual combination bridge, built in the 1870s, had a center swing span to allow the passage of maritime traffic, as well as a single Howe truss covered span and timber pile trestles at either end. The last example of a swing bridge in New Hampshire was the Squamscott River Bridge between Newfields and Stratham. This bridge was built in 1925 to replace a wooden swing bridge built in 1807 and was constructed by the Phoenix Bridge Company. It originally had a wooden deck which was replaced in 1947 with an open steel grid deck. The bridge soon thereafter became known locally as "the singing bridge" because of the high-pitched whine that resulted from rubber tires passing over it. This swing bridge remained in use until 1955, when the bridge was welded shut and was transformed into a fixed structure. Though no longer a moveable bridge, the singing nature of the bridge remained for nearly 50 years, a feature I remember well on my frequent travels over the bridge when I resided in the area! But, alas, the 1925 bridge has now been replaced; however one portion of the bridge, the mechanism used to rotate the bridge, still remains and is on display adjacent to the bridge.

Today, only the state of Maine has swing bridges remaining in its inventory of historic bridges. While at least eight metal swing bridges were built in the state from 1901 to 1954, only three remain. The most historic of these bridges are the 1931-built Kennebec River Bridge in Richmond, and the 1939-built Townsend Gut Bridge in Southport. Of the two bridges, the largest, and the most impressive, is the Kennebec River Bridge, though it seldom "swings" today. The Townsend Gut Bridge, on the other hand, is the busiest and perhaps the one with the most interesting history when it comes to its operators. Both will be discussed at further length in a subsequent chapter.

The final type of moveable bridge to be found in the region is the vertical lift bridge. As its name suggests, the vertical lift bridge moves a span of a bridge directly over a waterway and is usually distinguished in its appearance by the lift towers located at each end of the moveable span (though some lift bridges were later developed that did not require towers). Fixed spans on either side of the vertical lift span of the bridge, manned just as is a swing or bascule bridge, usually consist of Warren or Pratt truss forms. With its massive size and complicated lifting machinery, the vertical lift bridge was easily the most impressive and technologically advanced of all the moveable bridges. As with other moveable bridges, vertical lift bridges had been conceived and smaller examples built prior to the late 19th century. A number of impressive vertical lift bridges were designed in Europe, but were never built; and our old friend, Squire Whipple, built a number of small timber vertical lift canal bridges in New York during the 1860s. However, the modern development of this bridge not only came about with advances in steel construction in the 1890s, but, unlike the bascule and swing bridge types, was a recent invention pioneered by one of North America's greatest engineers and bridge builders of all time, J. A. L. Waddell. A native of Canada and educated at Rensselaer Polytechnic Institute in Troy, New York, as a civil engineer, Waddell was already a famed bridge engineer, author, and educator when he built his first vertical lift crossing, the South Halstead Street Bridge in Chicago, in 1894. Holding the patent rights to the vertical lift type of moveable bridge, Waddell, and his later partner John Harrington, would go on to build a number of famed examples of this type of bridge, as well as many other notable bridges in different forms, prior to his death in 1938. Among them were the Armour, Swift, and Burlington Bridge in Kansas City (where he founded his first consulting engineering firm in 1887), built in 1912; the 1926-built Bay Bridge in Newark, New Jersey; and the Marine Parkway Bridge in New York, built in 1937 and one of the longest vertical lift bridges in the world even today.

In northern New England, three vertical lift bridges have been built and all, constructed between 1921 and 1940, are still standing. Vermont has had no examples of this bridge type, New Hampshire and Maine share two of them, while the third is located in Maine. The oldest of these bridges is the 1921-built Memorial Bridge between Portsmouth, New Hampshire, and Kittery, Maine, which was engineered by Waddell. Next in age is the Carleton Bridge, built in 1926 between Bath and Woolwich, Maine, also designed by Waddell, and erected by the American Bridge Company. The last to be built was the Sarah Mildred Long Bridge, which also connects Portsmouth and Kittery. This bridge, completed after Waddell's death, was engineered by a former student of his who went on to become a well-respected engineer specializing in moveable bridges. All of these bridges are in operation today, though the Carleton Bridge is now bypassed for highway use and serves only railroad traffic. Its future remains uncertain. The other two interstate bridges in Portsmouth and Kittery are both vital links in the local transportation network and unique local landmarks.

5

The Life of a Bridge

Like the men that created them, most historic bridges have had a life of their own and, indeed, sometimes seem to be imbued with human qualities. As with other structures that play an important part in our everyday lives, such as our own homes and places of worship, bridges, too, are "born," have varied and ofttimes interesting and useful lives, and they "die." Sometimes the lives they lead, useful though they may be, are rather anonymous and unspectacular; at other times, bridges are born with great fanfare and live very public lives, never too far from the spotlight. As far as the death of a bridge is concerned, this too occurs under circumstances that are often mirrored that in human life. Sometimes, old bridges just gradually fade away, other times they suffer a violent or spectacular end, while in yet other circumstances, bridges die at a young age, their full potential never realized. But, what about the "lives" of the bridges which are the subject of our study, how, and under what circumstances, did they come to be built, and how did this process change over the years?

Planning, Purchase, and Safety Concerns

Early iron bridges, and later steel ones, came into being in northern New England, and elsewhere, in one of several situations. The most common of these by far was as replacements for earlier bridges that stood, or once did, in the same location, constructed either solely of timber, or ofttimes Howe truss bridges made of both iron and timber. Not surprisingly, as many as seven out of every ten such iron and steel bridges were built as more modern replacement bridges. However, in a few cases early metal bridges were built in new locations where no previous bridge had existed, and some were built as replacement bridges, not in the same location, but in a new location either further up or downstream in a location deemed to be more advantageous in one way or another. Finally, in some cases, new metal bridges were built in places where wooden bridges were not feasible, ofttimes replacing a ferry service. Local politics and preferences of land and business owners, perhaps more so than today, sometimes played a part in determining where a new metal bridge would be placed, as is evidenced by the Meadows Bridge in Shelburne, New Hampshire.

Whatever the situation, the decision to build a highway bridge was made almost exclusively at the town or city level in the northern New England prior to the 1920s, and it was in the locale in which it was built that the decisions as to what kind of bridge would be built and who would build it were made. While it is true, as with the iron bridge built at Highgate Falls, Vermont, in 1887, that state monies were sometimes granted to help pay for some bridges, the details were generally left up to the local community. Indeed, this is how the process had worked since colonial days. The early wooden bridges in the region were usually simple affairs that were easily built with local know-how by town officials or private bridge companies. However, with early metal structures, local selectmen (and some private toll bridge companies) were now tasked with overseeing the process of building a modern, scientifically designed structure about which, in most cases, they had no knowledge or experience. Few of the men in local government knew little, if anything, about iron bridges and the wide variety of trusses that were available, nor would they have been well-versed in the technical details of the process. In discussing the idea of bridge safety in 1887, one engineer, George Vose, a professor in civil engineering at Bowdoin College in Maine, stated the situation quite clearly when he wrote that "the ordinary county commissioner or selectman considers himself amply competent to contract for a bridge of wood or iron, though he may never have given a single day of thought to the matter before his appointment to office" (Vose, pg. 16).

So who, then, did local officials turn to for help and advice when they needed a bridge built in their town? It was primarily the bridge builders themselves. Prior to 1900, any bridge fabricator of any size or importance, companies such as the Groton Bridge Company, the King Iron Bridge Company, the Canton Bridge Company, the Berlin Iron Bridge Company, and the Wrought Iron Bridge Company, to name those most prominent in New England and the eastern states, employed traveling salesmen/engineers to sell their company's wares. These men went from town to town, city to city to sell their company's patented truss or arch designs and gained contracts to build bridges using both the old-fashioned method of pavement pounding and word-of-mouth recommendations, as well as more modern and sophisticated means such as mail order catalogs. Surprisingly, a simple "catalog" bridge could be bought by any community for any crossing.... All officials had to do to order one was to provide the simplest of details in regards to desired width, length, and like specifications! "Catalog" bridges, those ordered from a manufacturer and erected by local builders, were, however, rare for all but the smallest of crossings. One possible catalog bridge that is still in existence is the former Tannery Bridge of St. Albans, Maine, now on display in Pittsfield.

Those traveling bridge salesmen/engineers employed by the reputable companies previously mentioned were, by and large, selling tried and true structures to the towns they visited, even if it was their own patented design. Bridge salesmen were also noted, on occasion, for trying to sell town officials more than one bridge at a time. It seems likely, though there are no surviving records to tell us for certain, that a discount may have been offered if multiple structures were contracted for. In some towns, however, where more than one bridge was built by the same company within a relatively short time span, structures were sometimes purchased separately by both the local government and private citizens. This was the case in Franconia, New Hampshire, where the town purchased one lenticular truss bridge from the Berlin Iron Bridge Company, while a local farmer also purchased a similar sized crossing from the same company so that he could better access his own property. While I've found no surviving contracts between bridge builders and town governments from this early period, we may be sure that town selectmen, in keeping with the well-known New England traits of attention to detail, succinctness, and frugality, specified exactly what they wanted, and what they were willing to pay with in the contracts they signed with any bridge company.

However, two aspects of the bridge procurement process from the days of old may seem foreign to us in the modern age. First, the idea of putting a bridge contract up for bid and accepting the lowest bid was not yet standard practice. While most assuredly, local officials sought the best price they could get from a bridge company, we can also be sure that some of these companies may have greased the palm, so to speak, of a local selectman in order to gain a favorable vote or decision on who would be selected to build a bridge in their town. Secondly, another major consideration by local officials was the final price they would have to pay for their bridge. Then, as sometimes occurs even today, selectmen might be lured by the bargain price of a bridge company offer, without taking into account whether the bridge type offered was the best one suitable for its needs in terms of durability and safety, or without checking into the bridge company's background to see if it was a legitimate company. Commenting on this very issue in 1887, engineer George Vose wrote that "all over the country a great number of highway bridges which have been sold by dishonest builders to ignorant officials ... are on the eve of falling, and await only an extra large crowd of people, a company of soldiers, a procession, or something of the sort, to break down" (Vose, pg. 16). In talking about New England specifically, he further comments, "Not many years ago, a new highway bridge of iron was to be made over one of the principal rivers of New England.... The bridge, however, was a cheap one, and, as such, commended itself to the commissioners" (*ibid.*, pp. 16–17), despite the fact that an engineer had examined the plans and told town officials that the design was defective. The bridge to which Vose was likely referring was one built in his own state of Maine, a King Iron Bridge Company (KIB) built structure crossing the Androscoggin River between Brunswick and Topsham. Though Vose's concerns were valid, as another KIB-built iron bridge in Groveland, Massachusetts, had collapsed of its own weight in 1881, the bridge in Maine stood until 1907.

Just how often cost was taken into account by local officials over matters of reliability and safety in northern New England when it came to bridge building is unknown, but we may be sure that somewhere, at some time a bad decision was made, even if the bridge in question never failed in catastrophic circumstances. While major bridge disasters in northern New England attributable to faulty design or construction are non-existent, the short life span of some early iron bridges may be an indicator of either a poorly designed or built bridge, or one that was of a design unsuitable for a particular location. It would not be until the 1890s that the idea of hiring an independent consulting engineer, trained in the art of bridge building, to oversee the process during all its phases from bridge fabrication to final construction would become standard.

Bridge Fabrication and Construction

Once a bridge was contracted for in a given locale, the bridge contractor would soon begin to fabricate the bridge at its manufacturing facility. All but one of these bridge fabrication companies, the Vermont Construction Company of St. Albans, were located outside Maine, New Hampshire, and Vermont. Once a bridge was fabricated at the bridge company's plant, its components and fittings were all gathered and often numbered for ease of construction and were subsequently shipped to the final erection site. In the early days, these bridge components were primarily shipped by rail to the site where the bridge would be built, with perhaps the few remaining miles from a rail depot being covered by horse and oxen transport. Later on, rail and truck transportation were generally used. Of course, a bridge needed its supporting masonry work, abutments and piers when utilized, built before the bridge itself could be erected. This work was often contracted for separately by a given town and often employed local labor early on, though in later years it was often the work of the bridge company itself, or subcontractors from out of state with whom they had close ties.

Bridge construction in terms of the time it took to erect a new span covered a wide range.... Simple bridges might take as few as several weeks to a month, while large scale projects might take anywhere from six months to two years once actual construction began, though delays might be experienced due to poor weather, accidents, and other unforeseen difficulties. Such problems and delays have traditionally been attendant to public works projects in general in America, and the complaints that we have today about the numerous delays in road and bridge projects might well resonate with the American public in years gone by.

One aspect of early iron bridge building that was vastly different than today's bridge building activities was the lack of large cost overruns. With few exceptions, in the early days of metal bridge building, a project was proposed, a contract price was agreed upon, and the bridge was indeed built for that price. Cost overruns did occur at times, but were generally within reasonable limits, and the idea that any proposed bridge might cost one and a half times, or double the cost of the original contract price, as often happens today, was simply not acceptable. Two examples will suffice to illustrate this point. One of the state of Maine's first large bridge projects in the 20th century was the building of the International Bridge over the St. John River between Van Buren, Maine, and St. Leonards, New Brunswick, Canada, in 1909 and 1910. The amount of the original appropriations for the bridge, based on bids received from all parties involved for bridge fabrication, masonry work, road work and approaches, and consulting engineers, was $75,000. The final price came in at just under $85,000, primarily due to extra work that was not foreseen or planned for in the building of the piers and abutments. Even with this difficulty, the final cost overrun was less than fifteen percent of the total amount. The amount might have been more, but local businessmen in Canada and Maine, knowing of the cost overruns in the masonry work on the substructure, joined together to raise the money to pay for the approach fills to the bridge, and in turn asked the governments to reimburse them for only half the cost, which was gratefully agreed upon. This decreased the cost of the bridge by nearly $6,000 to Maine and Canada and is an interesting example of private citizens, who would benefit greatly from the project, working with the government to get things done close to budget and on time. In Vermont, too, we may find an example, albeit lesser in scale, of the exacting nature of bridge accounting and the efforts expended in bringing a bridge project in under or at the original budgeted amount. In 1884 the town of Highgate applied to the state for assistance in building Keyes Bridge at Highgate Falls. The state agreed to pay three-fourths of the cost of the bridge, which eventually ran to $14,000, some $500 over the original price. However, upon auditing the bills submitted by the town, the state auditor dismissed $50 worth of legal expenses as inappropriate, and $600 of the $4,441 masonry bill because the stone in the old abutments for the previous bridge could have been used in the construction of the new bridge! While there was yet another dispute about the pay for the commissioner who oversaw the building of this bridge in the amount of $400, in the end hawkish accounting resulted in a very low cost overrun, about five percent, for the Highgate Falls Bridge.

Construction of a bridge could take place at almost anytime during the year in northern New England, depending on the circumstances. Depending on how far along construction had progressed, even in the months of January and February some work could be done, as long as the bridge abutments and piers, if needed, were completed before the water over which the bridge stood became frozen. One interesting example is the 1930-built International Bridge over the St. John River at Fort Kent. Though conditions while working on the bridge in January 1930 were surely frigid and trying, the falsework does not appear to have been sunk in the river bed, and, set directly on the ice, was probably easier to set-up. Ideal construction times, as with today, were generally in the summer or early fall, when rivers and streams were usually at their lowest levels. However, this could change based on the weather conditions in any given year, as those of us who live in this region know all too well! Late winter and early spring, when snow melt and seasonal rainfalls could turn any river or stream from a placid beauty to a raging beast usually meant that bridge projects were planned to avoid these expected conditions. However, when floods hit the area (more about this later) and bridges were washed out, reconstruction plans sometimes proceeded at a rapid pace to replace an important lost link in the transportation network.

No matter when they were constructed, the building of these early iron and steel bridges in most any town was a major event. A combination of local labor and "imported" workers was used to build these structures. Local men were usually employed in grading and road work on the approaches to a bridge, as well as the masonry work for the average bridge. They were also sometimes employed in hauling bridge materials to the site, as well as providing

In this view the trusses of the International Bridge at Fort Kent are nearly complete. Note that the center truss panel at right does not yet have its top chord in place (courtesy Maine DOT).

local timber for deck planking or falsework, if required. On the other hand, a large or complicated bridge often required specialized masonry work supplied by experts outside the region who were used to working with the major bridge companies. As far as the erection of the bridge structure itself, this was done by the bridge crews employed by the bridge company itself. These men worked in teams throughout the region and, when they came to a particular town to build a bridge, were housed in local boardinghouses, hotels, or took rooms in private residences. Thus it was that the coming of the bridge crew was not only an economic boom to the town, but also provided a spark of excitement and expectation, and likely some gossip as well for local townsfolk. We may imagine that the boys in town were fascinated with the bridge construction and were surely thrilled to see the huge components raised into place. And could it be that a local lass or two may have had a short, innocent romance or flirtation with a handsome, well-traveled bridge worker in what little spare time he may have had? Of course, news accounts of the day seldom touch on these matters! The size of the bridge crew, led by a foreman or supervisor, could vary widely, of course, based on the size of the bridge to be built—perhaps from four to ten men, and even more if the project was a large one. It was the foreman who interacted with local town officials to keep them apprised of progress on the bridge. Construction methods for each bridge, again depending on overall length and number of spans needed, varied, but nonetheless followed established standards—scaffolding and falsework was generally required for all but the smallest structures. For small crossings truss components were often assembled on a river or stream bank and then hauled into place. Early on, simple pile drivers to erect falsework were used, as well as manual derricks to lift components into place. However, it was not long before bridge builders turned to power machinery to aid in these tasks.

When the trusses were in place, and the structure's deck was completed, the bridge was finally ready to open. Bridge opening ceremonies back then were important affairs, marking the start of a new era of transportation, perhaps, in a given town. In keeping with this important event, the newly completed bridge was often draped with the American flag and red, white, and blue bunting, and a ribbon-cutting ceremony held with local officials to mark the formal opening of the bridge. Ofttimes, parades were organized to pass over the bridge to formally mark the beginning of its service; or local town folk, dressed in their Sunday best, took a stroll upon the bridge. It is ironic that this was a time, when a bridge was at its newest and experienced large volumes of traffic at one time that might never happen again, that the bridge received its greatest test of strength, perhaps with the bridge engineer or bridge company representative looking on with some anxiety. While in some cases bridge collapses or partial failure did occur on these opening day ceremonies elsewhere in America, New England seems to have been spared such tragic events.

In regards to the above issues of bridge procurement, safety, financing, and construction, the decision making process gradually shifted over the years from the local level to the state level by the 1930s, when state highway and bridge departments matured and grew as the transportation network expanded in the region. Legislative action in the 1930s formalized these changes. In 1933 in New Hampshire, for example, the State Highway Department took full control of all trunk line road and bridge construction and maintenance, except for those roads and bridges within compact parts of a city with over 2,500 residents. In this case, the city or town remained responsible for the local infrastructure. Illustrative of these changes over the years are the bridge building activities of Concord, New Hampshire. How the iron and steel bridges of Concord from 1872 to 1933 came to be built was typical of the region as a whole in many aspects, but the details regarding these activities are quite interesting and clearly demonstrates how these changes occurred.

A Case Study: The Bridges of Concord, New Hampshire

The city of Concord, New Hampshire's capital, is situated in the south-central part of the state. Within its northwest environs are the headwaters of the meandering Contoocook River, while the Merrimack River slices through the center of the city its entire distance from north to south for about ten miles, including one major horseshoe bend that takes an eastward course before the river resumes its southern direction. Yet another river, the smaller Soucook River, has its course in the southeastern part of the city, serving as the borderline with neighboring Pembroke. First settled in 1727 as Pennacook Plantation, bridges have always been an important part of its transportation network. In addition to seven major bridges, the city had numerous smaller crossings by the 1870s. Except for the northern part of the city, the Fisherville area, waterpower for manufacturing concerns, while available in abundant supply, was vastly underutilized. Concord instead made its name as a rail and transportation center, in addition to the retail business generated by its position as state capital. Here it was that the stagecoach that won the west, the Concord Coach, was manufactured by the Abbott, Downing Company. But, more importantly, it was also where four railroads were also centered, making it one of the largest rail hubs in all of New England. The first iron bridge in the city, and probably the earliest iron highway bridge built in the state, was the Federal Bridge over the Merrimack River, which was contracted for when the old covered bridge was carried away by ice on April 11, 1872. Appropriations for a new bridge were made by the town within weeks in the initial amount of $10,000, and within six months a new bridge was contracted for with the Wrought Iron Bridge Company (WIB) of Canton, Ohio. It is possible other companies were contacted to provide estimates for a bridge, which would turn out to be a 3-span structure measuring 450 feet long, but there is no record of such dealings. The new bridge, a wrought iron and channel column arch structure, was completed and opened for public travel on April 1, 1873: the bridge structure's final cost was $18,463, while the abutments and stone work cost an additional $8,500. Of this amount, $45 was paid to an engineer, Charles C. Lund, who was an early civil engineer in the state. Though the work paid for is not specified, perhaps it was for reviewing the plans of WIB, or possibly for work on the abutments. As to the abutments, all the granite was obtained locally from the quarry of Luther Roby at Rattlesnake Hill near an area called the "Pulpit." Indeed, the granite industry was a growing concern in Concord by 1850 and provided not only the building material for its capital buildings, but also the stone for the abutments of city bridges built well into the 20th century.

In 1874, two more bridges were contracted for by the city; the Fisherville Bridge over the Merrimack River, a double intersection Whipple truss structure, was also built by the Wrought Iron Bridge Company (WIB) to replace a covered bridge, with no record of any other bids being received, and cost a total of just over $17,000. Of this amount, $11,701 was paid to WIB for the iron bridge, and the rest for stone, cement, and local labor. No independent engineer was paid for any aspect of the work. Interestingly, that same year the town also built the Sewall's Falls Bridge, a wooden covered bridge over the Merrimack, paying Dutton Woods $14,411 for the job. The bridges built by WIB apparently served the city well for some time, but in 1898 some trepidation was felt by the town fathers, who thought the Fisherville Bridge (later called the Main Street Bridge) might be too lightly built. Accordingly, they sent a letter to WIB asking for one of their agents to thoroughly inspect the bridge. Interestingly, while the safety of the bridge was being questioned, the inspection was requested from WIB because "it was considered that they would give a more favorable report of its condition, not wishing to condemn their own work" (*Annual Report of the City of Concord, 1898*, pg. 205). Despite this incredible assumption, the town was sadly mistaken; in March 1898, the report on the condition of the bridge was received, and the news was not good. The verdict rendered by WIB's eastern agent, James Wynkoop, was emphatic; "We deem it our duty to warn you that the bridge is unsafe and a menace to life and property. It should be closed at once to all heavy traffic and light vehicles should be required to proceed slowly when crossing it. Care should also be taken not to allow many cattle or sheep to cross in a bunch and crowds of people should not be allowed to gather on the bridge" (*Ibid*, pg. 208). Despite fishing for a favorable report on the condition of the Fisherville Bridge, the town followed Wynkoop's advice and replaced the bridge with another bridge provided by WIB (cost and type not specified). However, the old Fisherville iron bridge was not scrapped, but was instead shortened and erected in place of a condemned covered wooden span of the Twin Bridge over the Contoocook River in Concord. With this work, as well as the building of a small iron bridge on the Dunbarton Road at St. Paul's School, Concord's bridges, a mixture of modern iron bridges and covered bridges, were deemed in good shape for years to come. However, less than two decades later, in 1915, five of the cities largest bridges were once again in need of replacement. This time, at least four of the bridges were put up for bid, with ten bridge companies competing, and every contract was awarded to the lowest bidder. The Pembroke, Borough, Federal, and Sewall's Falls bridges, all truss bridges, were awarded to the Berlin Construction Company, while the Main Street (formerly Fisherville) Bridge, a plate girder bridge, was awarded to the Penn Bridge Company. The Pembroke (or South End) Bridge was formerly a covered bridge, replaced by a four-span structure consisting of two high Pratt and two pony Warren trusses. Coming in at a total cost of $25,508, this was the most expensive of the five bridges, its completion delayed by unusually heavy rain for twenty-eight days. The Borough Bridge, a much smaller structure crossing the Contoocook River in the northwest part of the city, cost $4,526 and encountered no difficulties in its construction. The Federal Bridge, on the other hand, cost $24,451 and was delayed in its completion by high water carrying away the false-works for one entire span, and part of another. Extra expenses were also incurred on this bridge with the addition of a sidewalk not in the original plans. Finally, the Sewall's Falls Bridge, which took the longest time to build (perhaps because the same construction company was used to build all the new bridges, as no mishaps were otherwise reported), cost $14,227 and was the only one of these five bridges to retain a wooden floor. The costs for all these bridges were slightly offset by the sale of scrap metal and wood from the old bridges, not an uncommon practice and one entirely in character with New England's frugal traditions. The sale of cast iron (probably sidewalk railing components) from the old Federal Bridge brought in a paltry $25, while the wrought iron from the old Fisherville and Federal bridge superstructures fetched $748. In contrast, the old lumber from the Pembroke Covered Bridge brought in $541.

It is interesting to note that while all these 1915 bridges were constructed by the Daniel Marr and Son Company of Boston under the direction of foreman Harry Martin, local civil engineers John and Edward Storrs were paid a total of $3,927 for designing all five bridges. It is significant to note that John Storrs at this same time was employed by the state of New Hampshire on its Public Service Commission, which was in charge of the state's public roads, bridges, and railroads. Though his private engineering jobs would seem to be a conflict of interest with his public role in regards to modern appropriation and building policies, in 1915 this was not the case and it is highly unlikely that few civil engineers in the state, or probably the region for that matter, had more experience at building bridges than John Storrs.

The final transition from local to state control of bridge building activities (along with increasing federal aid) in New Hampshire is illustrated by the building of the new Manchester Street Bridge (formerly the Pembroke Bridge) eighteen years later in 1933. The bridge built in 1915 came under state control in 1933 due to its location on the Daniel Webster Highway (U.S. Route 3) and was a good candidate for replacement and relocation due to safety concerns with heavy traffic and the need for a straighter highway and realigned approaches. The new bridge was a three span steel combination structure, employing a high Parker and two high Pratt trusses. The bridge was fabricated by the Lackawanna Steel Construction Company, but was designed locally by New Hampshire Highway Department engineer, Sheldon Hare, under the supervision of the state's assistant bridge engineer, Harold Langley. The cost of the Manchester Street Bridge is truly indicative of how much things had changed in less than twenty years. Whereas the city of Concord bought five new bridges in 1915 for about $80,000, this single new bridge cost $109,376, including a $10,000 charge for removal of the old bridge (the spans of which were subsequently reused elsewhere in the state in Henniker and Tamworth). Of this amount, Concord contributed $8,000, as well as the land on which it rested. However, we may be sure that the bridge was strong and worth every penny of its cost, as at least 5,000 people, and possibly as many as 7,000, attended its dedication ceremony on August 11, 1933. This bridge lasted for 64 years before being replaced and, at the time of its removal, was the longest through truss bridge in the state. Yes, bridge construction in Concord had certainly changed over the years!

6

Civil Engineering in Northern New England

Competition among early bridge building companies on the East Coast was fierce, but the landscape changed dramatically in the United States when financier J.P. Morgan purchased twenty-eight well-known independent bridge companies in one fell swoop in 1900, thus creating the American Bridge Company (Ambridge). The following year, Ambridge became a subsidiary of the U.S. Steel Corporation, yet another J.P. Morgan creation. While a number of independent bridge companies survived this consolidation and takeover by Morgan, after this time bridge companies were increasingly associated, not surprisingly, with major steel manufacturers. A list of all the known bridge companies that were active in northern New England may be found in Appendix 1. However, though the story of the trusses developed by these companies has previously been related, the location of these companies, save one, outside northern New England makes a full account of their activities less germane to this story. But, what about the state and local engineers involved in designing these structures? Maine, New Hampshire, and Vermont each had state and local engineers, as well as educational institutions that played an important part in the development and construction of iron and steel bridges, while several outside engineering firms which designed important bridges in northern New England are also worth highlighting. It should, however, be noted that when it comes to the state engineers discussed below, these names are but a small sampling of the engineers at work, the leaders of departments that at any given time also employed as many as two or three dozen assistant engineers, draftsmen, and other personnel. These were men that contributed greatly to the overall process of getting a bridge designed and built, even if their names or the details of their careers are less well documented.

Maine

The state of Maine has received but scant attention in regards to its metal bridges when compared to the other New England states. This is largely due to the fact that many of the older pioneer metal bridges, including those of the lenticular and multiple intersection Warren or Whipple trusses, were aggressively replaced in the 1920s and 1930s by more modern structures that, while cost effective and long lasting, utilized truss types that had become old standards by then and, thus, received little notice. Additionally, the "heroic" long span bridges in the state, such as the Steinman-built suspension bridges at Prospect and Deer Isle and the vertical lift Carleton Bridge at Bath were all engineered by those outside the state of Maine, and were dwarfed by larger bridges of the same type elsewhere in the United States. Despite Maine's lack of attention, its early engineers were well regarded for their skill and leadership.

Even before the advent of highway departments and any developed standards in bridge engineering, native engineer George Vose was a leader in campaigning for reforms in the methodology and business of building bridges. Born in Augusta in 1831, Vose was a professor at Bowdoin College from 1871 to 1881, authored a bridge engineering textbook, and by 1902 was a professor in engineering at the Massachusetts Institute of Technology (MIT). His condemnation of bridge building and inspection practices was directed not just at bridges in Maine, including a railroad bridge over the Hampden River near Bangor, but nationwide. A scathing letter he wrote to a Portland, Maine, newspaper was even reprinted in full in the *New York Times* with the headline "Bridge Inspection in Maine. What a Bowdoin College Professor Thinks of the System — A Bridge Officially Condemned Four Years in Succession and Still Used" (*New York Times*, February 25, 1878, n.p.)

The first of Maine's state engineers to come to real prominence was Paul D. Sargent. A native of Machias, he graduated with a civil engineering degree from the University of Maine in 1896 and saw his first work with the Washington Railroad as assistant engineer from 1897 to 1903. In 1905 he was appointed as the state's first Commissioner of Highways. Initially focused on roads, Sargent would soon turn to the subject of bridge building, providing the plans for the 1908-built steel swing bridge built in his own hometown between Machiasport and East Machias. Sargent's greatest gift to Maine, perhaps, was his activities in "placing highway bridge building on a business-like basis with fair and competitive bidding" (*Highway Bridge Building in Maine*, pg. 6). Sargent's ideas, however, were slow in being accepted and clashes with officials at the local level, who sometimes looked solely at cost and not quality of design, were not uncommon. Illustrative of this was Sargent's design for a concrete bridge abutment for a structure in Whitneyville across the Machias River in 1907. It was rejected by local officials in favor of a cheaper, poorly constructed wooden abutment instead. This type of dispute, and the political infighting involved in Sargent's position caused him to resign (though he may have been forced out of office) in 1911. He subsequently left Maine to work for the federal Office of Public Roads as assistant director, but returned to office in Maine in 1913 when the Maine State Highway Commission was formed to replace the Commissioner of Highways. Appointed as chief engineer with everyday charge of operations, Sargent and the State Highway Commission, in an unusual move, declared that the laws under which they operated did not include bridge work, only highways, and that local towns were responsible for bridges over 12 feet in length! Did Paul Sargent and the commissioners really believe in this arrangement? Certainly not, for they "had correctly gauged that the state legislature would vote in favor of increased funding and a more systematic and business-like approach to bridge construction and maintenance" (*Ibid*, pg. 9). Sargent, surely wary of the many disputes which arose during his previous tenure, knew that things had to change, and they did. In 1915 the Maine legislature passed the General Bridge Act with regular appropriations and plan review by the highway commission. The commission, as a result of this act, planned on organizing a bridge department that would be headed by an engineer qualified to build any type of highway bridge, but these plans seem to have been delayed by World War I, a time during which there was a lack of labor and funding for bridge projects anyway. The service of Paul Sargent to the state of Maine, though largely forgotten today, was extremely important in bringing the state into the modern era and through its first growing pains.

Sargent's successor in Maine, and its first state bridge engineer, was Otisfield, Maine, native Llewellyn N. Edwards. Born in 1873, he, like Sargent, attended the University of Maine and graduated with a degree in civil engineering in 1899, subsequently gaining his masters in 1901. His early positions in business were significant, and gained him much valuable practical and business experience. He first served as a draftsman with the Boston Bridge Works (1901–1902), and subsequently worked in the bridge design departments of the Boston and Maine Railroad (1903–1905), The Chicago and Northwestern Railway (1906–1907), and the Grand Trunk Railroad (1907–1912), for which he engineered five bridges on its Lewiston Branch in Maine. From here, Edwards turned to the public sector, serving as bridge engineer for the city of Toronto, Canada, from 1913 to 1919, and thence returning to the U.S. to work as a senior bridge engineer for the Bureau of Public Roads in Arkansas, Texas, Oklahoma, and Louisiana. During this time, Edwards published a pamphlet on practical bridge foundation design and stressed the need for individual design solutions even when using standard technologies and methods of building. This philosophy and attention to detail "would become a hallmark" (*Ibid*, pg. 12) of Edwards during his leadership of the Bridge Division of the Maine State Highway Commission, a position to which he was appointed in 1921. During his subsequent tenure from 1921 to 1928, 330 bridges were built, the vast majority of which were concrete structures. Indeed, while Llewellyn Edwards built only fifteen steel truss bridges during his time, he stressed the need for structures that were strong and had liberal allowances for stress overload, but also ones that were "pleasing to the eye" (*Ibid*, pg. 13). Few of Edwards' bridges remain today, the most significant being the 1921-built International Bridge at Madawaska. Upon leaving the Bridge Division in 1928, Edwards returned as a researcher for the Bureau of Public Roads in Washington, D.C. In later years he would write *The Evolution of Early American Bridges*, a significant but hard to find work that was published posthumously. Llewellyn Edwards retired from the BPR in 1943 and died in 1952, but his collection of works on early bridge engineering became "the nucleus of an important collection housed today at the Smithsonian Institute" where the "curator of engineering still uses Edwards old rolltop desk" (*Ibid*, pg. 17).

Our final Maine engineer of note is Max L. Wilder (1894–1962), born in Augusta, and, like Sargent and Edwards, a graduate of the University of Maine. Upon gaining his civil engineering degree in 1914, he was hired by the Maine State Highway Commission. Well respected and with thirteen years of experience (he served in World War I for two years) under his belt, he took over for Edwards in 1929. Wilder is noted for leading the Bridge Division during some very difficult and active times, including the Great Depression of 1929, President Roosevelt's New Deal Era, and the terrible flood year of 1936. While Wilder was considered conservative in his bridge designs, he nonetheless built tried and true structures that have served the state of Maine well for many years now. The most significant of these bridges are the Morse Street Bridge in Rumford, the seven span swing bridge in Richmond, the continuous truss bridges in Durham and Hollis, and the cantilever arch bridges at Arrowsic and Augusta, the former of which was renamed in his honor after his death in 1962.

New Hampshire

As previously discussed in the opening chapter, this state had a fine reputation in the field of wooden covered bridge building, one that would eventually carry over into the building of steel bridges as well by the state. Of the early engineers, there is little of record, as is the case with Maine and Vermont, and it seems likely that most early iron bridges were designed by the bridge companies themselves. One early name that has been found for New Hampshire is that of Charles Carroll Lund. He was a native of Concord and studied civil engineering at Dartmouth as a sophomore during the 1852–1853 school year and graduated ca. 1855, but his engineering activities during the early years of his career are unknown. In 1873 he was paid $45 for providing engineering services for the first iron bridge built by the city of Concord, the Federal Bridge, the only specific record I can find of his engineering work. However, his true importance may lay in the tutoring he provided in later years to an aspiring civil engineer, John W. Storrs.

When it comes to metal truss bridge engineers, John William Storrs was one of the most active men in the region at the height of the building boom for truss bridges, and is thus a significant figure. A native of Vermont, Storrs was born in Montpelier on November 24, 1858;

but at a young age his family moved to Concord, New Hampshire, where he attended the public schools. A weak child by his own account, Storrs suffered from an attack of scarlet fever at the age of ten. While the disease killed his sister, Storrs was left with reduced eyesight and was largely deaf. He was an avid reader and a good student in his youth, but none too athletically inclined. How or why John Storrs took an interest in civil engineering is unknown. He was a practicing civil engineer by 1885; but where he obtained his degree I've been unable to ascertain, though one brief biography states that he studied under Dartmouth alumni Charles C. Lund. The following year, he married his wife, Carrie, who would be his companion for the remainder of his life. Upon entering the field of civil engineering, John Storrs worked for a number of railroads, including the Concord Railroad, the New Boston Railroad, and for the Boston and Maine Railroad from 1890 to 1911. During this time he designed a number of metal truss railroad bridges, one of the oldest known remaining being that over the Pemigewasset River between Ashland and Bridgewater, built in 1903 by the American Bridge Company. Storrs, both alone, and later with his son Edward Storrs, was very active in designing bridges and consulting throughout the state and the region on jobs large and small. During his entire career as an engineer, John Storrs wore a wide variety of hats, sometimes all at once! While still employed with the Boston and Maine Railroad, Storrs became the state's first highway official, in charge of the state's main roads and bridges. During this time, as previously detailed, bridge building was usually controlled at the local level, and it was John Storrs in many cases that did the design work. I've already documented his activity in Concord, where he designed five bridges in 1915; one of which, the Sewall's Falls Bridge, is the only one of his structures remaining in his adopted hometown. Other extant Storrs designed structures include those in Milford, Wentworth, Henniker, and Hinsdale, as well as the former railroad bridge (now a highway bridge) over Quichee Gorge in Hartford, Vermont. Not only was Storrs an engineer, he was also a writer, authoring with his son and business partner, Edward Storrs, *A Handbook for the Use of Those Interested in the Construction of Short Span Bridges* in 1918. Storrs worked for the state of New Hampshire until 1930, when he was forced to retire due to his age. Though he loved moving pictures, reading, and card playing (bridge, naturally!), Storrs was not yet done with public life.... In 1933 he was elected the mayor of his hometown, having made no campaign speeches, and would go on to be elected to serve for five more two-year terms. Despite his salty language and bull-dog personality, Storrs made a real dent in the city of Concord's problems, cutting its debt by 1940 by over $250,000, eliminating personnel, and boosting the tax rate to support education. Upon being reelected in 1940, he contended, "I am the most bitterly hated man in Concord. That's why they elected me mayor. I won the city's unpopularity contest" (Gross). Despite this hyperbole, Storrs, said to be New England's oldest mayor in 1940 at the age of 82, truly enjoyed his job. When he died in 1942 he left behind a truly unique legacy of both public structures and public service.

Before leaving the private sector, one other civil engineer from New Hampshire, William Mann, may also be mentioned, though how much building he actually did in the state is unknown. He was a graduate of Dartmouth in 1893, where he subsequently gained his masters degree in 1896. Mann worked down south for the Mississippi River Commission in 1893, the Berlin Iron Bridge Company in Connecticut in 1895, the New York, New Haven, and Hartford Railroad in 1896, the Boston and Albany Railroad from 1897 to 1898, and the Rutland Canadian Railroad from 1899 to 1901. After this time, Mann was the junior partner in Lloyd and Mann, Civil Engineers in Concord, New Hampshire. Whether he built any bridges in his native state is unknown, though his tenure with the Berlin Iron Bridge Company may have resulted in some work somewhere in northern New England.

Information on the bridge engineers of the New Hampshire Highway Department is sadly incomplete. However, through the research of state historians James Garvin and Richard

Casella, a number of men and their careers have been brought to light. All are notable for the bridges they designed, and some for their additional work beyond the state level. One of the earliest of the state engineers was John Warren Childs, a resident of Merrimack, New Hampshire, born in 1888 and a graduate of Dartmouth in 1909. Though little is known of Childs' family background, it is possible — indeed likely — that he was directly related to the Childs' brothers, Horace, Warren, or Enoch, of the well-known family of New Hampshire bridge builders from Henniker. Whatever his background or inspiration, John W. Childs joined the New Hampshire Highway Department in 1916. Within six years he was the division engineer in Littleton, and by 1925 was appointed chief bridge engineer, a position he held until 1942. During this time period he oversaw the construction of hundreds of bridges of all types, indeed all but two of every single bridge built by the state, and provided professional leadership during a period of change, and, in the wake of the floods of 1927 and 1936, a flurry of bridge building activity.

The assistant bridge engineer for the state of New Hampshire during Child's tenure was another accomplished engineer, Harold E. Langley. Though he joined the department in the 1920s, virtually nothing is known of his family or educational background. He designed hundreds of bridges in all forms, both concrete and steel, and, though outside the scope of this work, was instrumental in introducing the concrete rigid frame as a standard bridge design in the state. However, Langley was also highly skilled at designing steel bridges, ranging from the simple and practical, such as the 1936-built Huntress Road Bridge in Freedom and the 1928-built Bridge Street Bridge in Littleton, to larger and more complex spans, such as the award winning, 1930-built Beecher Falls Road Bridge over the Connecticut River, a steel arch deck structure. In addition to his state bridge work, Langley was also active in training future civil engineers.... As an expert on long-span structures, he co-authored the second edition of George Hool and W. S. Kinne's standard text *Moveable and Long-Span Steel Bridges* in 1943, contributing the chapter on steel arch bridge design, in which he highlighted the recent steel arch bridges built over the Connecticut River by the state of New Hampshire at Chesterfield and Orford in 1937. These were both bridges that Langley worked on closely with his boss, John Wells.

Another man whose work was extremely important, and award winning, in New Hampshire bridge design was John H. Wells: he was a graduate of Worcester Polytechnic Institute (WPI) of Worcester, Massachusetts, in 1930 and soon after began working for the New Hampshire Highway Department. By 1935 his initials appear on bridge plans, and by 1937 to 1938, with his design of the steel arch bridges over the Connecticut River at Chesterfield and Orford, had proven himself a forward-thinking and highly skilled engineer. Wells would subsequently leave New Hampshire for a job in the private sector, working with the engineering firm of Jackson and Moreland in Boston by 1959. He would remain with this company for years, rising to the position of chief structural engineer by 1970. John Wells is also interesting in that his career spanned the gap between the old and modern times. When designing his arch bridges in New Hampshire, he made "dozens of pages of densely packed calculations" (Bartlett Bridge Survey, pg. 9) without benefit of a calculator, while in 1959 he worked with "the new IBM 650 electronic digital computer, which allowed engineers to design the structural steel for buildings with the aid of a mathematical program contained on 1,000 punched cards" (*Ibid*).

Other known bridge engineers for the state that worked in the hectic years of the 1930s, though this list is incomplete, includes the following; Henry B. Pratt, Jr. (no known relation to the Pratt truss inventors), a graduate of WPI who was employed from 1935 to 1937 and designed three bridges, including a non-extant Pratt truss bridge at Penacook and the Hancock-Greenfield Covered Bridge (essentially a covered Pratt truss structure); Sheldon T.

Hare, who designed two bridges in 1939 and 1940, now non-extant, at Bartlett and Canaan: F.W. Brown, who designed the 1937-built Oak Street Bridge, utilizing a high Warren truss, in Newport; E.B. Davis, designer of a non-extant high Pratt truss bridge erected in Errol in 1941; and Robert J. Prowse, born 1906, a native of Concord, New Hampshire, and graduate of Northeastern University in Boston, who worked briefly for the Hamilton Bridge Company of Ontario, Canada, before joining the New Hampshire Highway Department in 1933. He would design award winning bridges in the 1950s, become the state's chief bridge engineer in 1968 and, after his death in 1969, the Ash Street Bridge in Londonderry, which won an award in 1959, was renamed in his honor.

Vermont

This state is unique among the states of northern New England in three aspects of bridge engineering history. The first is the fact, based on the research of bridge historian Robert McCullough, that the state had employed its first bridge engineer, Charles H. Clark, in the 1880s, nearly twenty years before Maine and New Hampshire would hire their own experts. He is noted for designing two Lake Champlain bridges, the 1886-built North Hero-Alburg swing bridge, a five span riveted lattice truss structure measuring 850 feet in length. Three years later, Clark designed the Grand Isle-North Hero swing bridge, a single span structure with extensive causeways connecting it to each shoreline. Begun in 1889 and completed three years later, this structure was a toll bridge whose building process was marred by lawsuits over quality and cost issues. The state of Vermont, with Clark in charge, ultimately ended up paying for half the cost of the bridge.

Both of the bridges designed by Charles Clark were fabricated by the Vermont Construction Company (VCC) of St. Albans, which points to another unique aspect of bridge engineering in Vermont; it was the only state in northern New England which was home to a bridge fabrication company. The VCC had its roots from the Howe Bridge Works company established in 1840 in Springfield, Massachusetts, by the truss designer William Howe, and soon thereafter sold to his brother-in-law Amasa Stone and business partner D. L. Harris. One of this company's young employees was Richard F. Hawkins, who, though he did not have a civil engineering degree, was quite skilled and rose through the ranks to become a partner in the bridge company by 1862, and its sole owner a few years later. The R.F. Hawkins Iron Works, by now a very successful bridge fabricator in New England, purchased the rolling mills of the bankrupt St. Albans Iron and Steel Works company in Vermont in 1884 after being awarded the contract to build the North Hero-Alburg swing bridge, and incorporated the Vermont Construction Company as a subsidiary in 1886. The company articles list Hawkins as president, while it employed two brothers, Charles and Edward Babbitt, as superintendents. Charles Babbitt ran the iron division, his brother the wood division, while E.B. Jennings was consulting engineer. Hawkins' move to form the VCC was an astute one; the need for bridges in the Lake Champlain region was growing, and the VCC, with its fabrication facilities right on the scene, was able to capture much of the action, especially in the railroad business, throughout the late 1880s to the mid–1890s. However, not all of the VCC's work was localized. The company built bridges in New York and every New England state, except Rhode Island, including four bridges in Campton, New Hampshire, from 1896 to 1900 and a three-span girder bridge in Dover, New Hampshire. The Campton, New Hampshire, bridges built by the VCC were the result of unusual circumstances. By 1896, Edward Babbitt, one of the founding superintendents of the VCC and its wood division, had left the company to work for the Boston and Maine Railroad, and later was a resident and selectman of Campton. When the town needed some modern bridges in the late 1890s, it is not surprising, then, that Babbitt would not only use his old business or family connections (it is unknown if Charles Babbitt was still employed by the VCC at this time) to the town's advantage, but also acted as bridge engineer and supervisor during the erection of the Osgood Bridge, and perhaps the Bog Brook, Dole, and Route 175 (relocated in 1927 to Spencer Brook) bridges in Campton as well.

The VCC would only have a relatively short life, and by the last year of the century its business had declined. With J.P. Morgan's monopoly of the bridge business, spurred by the formation of the American Bridge Company in 1900, the VCC was nearly out of business. It was sold by its parent company, R.F. Hawkins, to a group of buyers from Vermont and New York in 1901 and was renamed New England Bridge Works. However, it seems unlikely the company ever built any bridges, and the rolling mills at St. Albans closed that same year. The company's charter was finally revoked in 1906, and the name taken by another corporation. The new VCC operated from 1906 to 1912 and included as one of its investors and operators civil engineer Frank Sinclair, former engineer for the city of Burlington, and at one time an agent for the Pittsburgh Bridge Company. Sinclair's company did build a few bridges, though their design types are unknown.

While the VCC was the only northern New England based bridge fabrication company during the era of the metal truss bridge, one other company with Vermont connections is also worthy of mention, that of the King Iron Bridge Company. Founded and based in Cleveland, Ohio, in the early 1860s, its founder was Vermont native Zenas King, born in Granville in 1818. As a young man, King made his way to southern Ohio by 1840, and in the late 1850s was a salesman for Thomas Moseley and his bridge company, which later relocated to Boston. Though not trained as an engineer, King patented his own version of a bowstring arch truss and went on to start his own company based in Cleveland. The King Iron Bridge Company was truly a family affair, with sons James and Harry heavily involved in its day to day activities. Among the few King bridges known to have been built in Vermont are the still extant Zenas King Memorial Bridge, a beam girder structure in his hometown of Granville, and one that once stood in Woodstock.

Yet another unique facet of bridge engineering in Vermont was the early establishment of an engineering program at a state educational institute. While New Hampshire and Maine had well established engineering programs at Dartmouth College and the University of Maine by the last half of the 1800s, Norwich University in Vermont had such a program as early as 1825. The University of Vermont would establish such a program at a much later time in the 1860s, with Professor Volney Barbour being the sole faculty of the engineering staff for many years. In 1884, another noted engineer, Josiah Votey, joined the engineering department at the university. He would later become one of several pioneering engineers employed by the state in the early 1900s to give free engineering consultations to towns in need of such advice. As historian Robert McCullough notes, both Barbour and Votey had very long tenures at the University of Vermont (UV), and taught several generations of young men the trade of civil engineering in the state.

By 1912, the Vermont Highway Commissioner's office, with approval from the legislature, began to supply local towns with bridge plans and oversee bridge building activities; and by 1917, with increased federal funding, bridge building activities increased. By 1921, the establishment of a state engineering department was fully realized. In discussing Vermont bridge engineers, the following men have also been identified by Robert McCullough as early state bridge engineers; Herbert McIntosh, an 1890 UV graduate, former engineer for the city of Burlington and later partner in his own engineering company; Hubert Sargent, named chief bridge engineer for the state in 1925 after many previous years of experience; George Bishop, a bridge engineer for the state for 35 years before becoming the chief engineer; George Reed, one of the first bridge engineers employed by the state during the 1916–1921 growth period;

and Jasper Draffin, a 1913 UV graduate in civil engineering, with a graduate degree from the University of Illinois in 1916, briefly a teacher at the University of Illinois, state bridge engineer for an unknown time, ca. 1920–1930. Once again, as with Maine and Vermont, this is by no means a comprehensive list of the bridge engineers working in the state.

Regional Engineers

Though most bridges built in Northern New England were homegrown, so to speak, there were a small number of long span bridges built in the region whose design and construction was superintended by outside engineering companies whose reputations were well-established nationwide.

J.A.L. Waddell and John L. Harrington

We have previously taken a look at John Alexander Low Waddell (1854–1938) and his background in regards to his development of the unique vertical lift type of moveable bridge. He operated his engineering firm under a variety of names over many years, including Waddell and Harrington (1905–1915), Waddell and Son (1916–1920), J.A.L. Waddell (1921–1926), and Waddell and Hardesty (1927–1945). The firm of Hardesty and Hanover continues to this day, and holds Waddell's original drawings for the 1921-built Memorial Bridge connecting Portsmouth, New Hampshire, and Kittery, Maine, and the 1926-built Carleton Bridge in Bath, Maine.

A former partner of Waddell's, and a man who held many patents with him in the continued development of the vertical lift bridge was John Lyle Harrington, an 1895 graduate of the University of Kansas. After splitting with Waddell in 1914, Harrington formed his own engineering partnerships over the years, the final version being Harrington and Cortelyou in 1928. The junior partner in this firm, still active today, was Frank Cortelyou, a long-time assistant engineer to Harrington. Together, they designed the 1940-built Sarah Mildred Long Bridge, a vertical lift structure connecting Maine and New Hampshire.

Joseph R. Worcester

J.R. Worcester (1860–1943) was a native of Waltham, Massachusetts, and a noted steel bridge engineer who designed and built many bridges from the 1890s through the 1930s. He graduated from Harvard University in 1882 and first worked for the well-known and prolific Boston Bridge Works (BBW) company, working his way up from junior draftsman to the company's chief engineer. In 1907 he left BBW to form his own company, J.R. Worcester and Company, working there until his retirement in 1924. He is noted for his work nationally, being appointed by President Herbert Hoover in 1921 to serve on a committee to develop national standards for building codes and materials. Closer to home, he designed most of the steel structures for the Boston Elevated Railway, and was noted for his arch bridges, designing the 1905-built steel arch bridge built over the Connecticut River at Walpole, New Hampshire, as well as the steel arch bridge over the Connecticut at Woodsville, New Hampshire. Worcester's firm later also designed the Notre Dame Bridge in Manchester, New Hampshire, another notable steel arch bridge. It is unknown if J.R. Worcester himself contributed to this project, though his son Thomas Worcester and the firm's chief engineer at this time, Charles Turner, were involved.

Fay, Spofford, and Thorndike

This engineering firm was established in Boston in 1914, and gained real prominence in the 1930s with the addition of Charles M. Spofford (1871–?). He had previously served in the

Civil Engineering and Environmental Department at the Massachusetts Institute of Technology (MIT) from 1911 to 1933, and was later the author of *Theory of Continuous Structures and Arches* in 1937. Two of Spofford's best known designs were for the still extant 1934-built General Sullivan Bridge in New Hampshire, a continuous truss steel bridge, and the historically important Lake Champlain Bridge, which served from 1929 to 2009. This bridge was significant as the first major continuous truss structure designed for highway use in America. One of this firm's other creations was the large, albeit less innovative, Rouses Point Bridge on Lake Champlain, a thirteen-span, Warren truss structure built in 1937 that served until the late 1980s.

David B. Steinman — America's Suspension Bridge Expert

Born in 1887, David B. Steinman was a native New Yorker who gained world renown as an innovator in the building of epic suspension bridge. Indeed, if ever there was a case where a future bridge builder's career seems to have been preordained, such was surely true of Steinman. The Brooklyn Bridge was completed just four years before his birth, and he would literally grow up in its shadow. A bright student, he took engineering courses at the Cooper Union Night School while still a young teenager and at the same time attended City College of New York — which had no engineering program — and gained a degree in teaching, qualified to teach vocational courses in iron and wood working. From 1906 to 1910, he attended Columbia University, graduating with a PhD in engineering. His doctoral thesis, which received a perfect mark, was written about suspension and cantilever bridges. Following his graduation, Steinman had a spectacular career as a teacher, bridge engineer and designer, and writer, which is well known and much too long to recount here. Among the early notable structures he designed outside northern New England were the Carquinez Strait Bridge in California, the Steubenville Bridge across the Ohio River, the Florianopolis Bridge in Brazil, the Mount Hope Bridge in Rhode Island, and the St. John's Bridge in Oregon. In 1929 he was invited to submit a proposal for a bridge across the Penobscot River in Maine and was subsequently awarded a contract as design engineer for the Waldo-Hancock Bridge. Built during the height of the Great Depression, this bridge cost a mere $800,000, coming in $400,000 under budget, and was among Steinman's personal favorites among all his creations. During the later years of the depression, when bridge funding was hard to come by, Steinman built the Deer Isle Bridge on the Maine coast, an economical and long-lasting structure spanning Eggemoggin Reach that he himself regarded as less aesthetically appealing than the Waldo-Hancock Bridge. In addition to his Maine bridges, David Steinman built a number of notable bridges between 1930 and 1958, his career culminating with the epic Mackinac Bridge across the Straits of Mackinac in Michigan.

7
Railroad Bridges

The importance of the railroads nationwide in expanding the use of iron and steel bridges has previously been mentioned. While there were many independent branch railroad lines in northern New England in the last half of the 19th century, by the 1890s many of these were consolidated into several large corporate networks, such as the Maine Central, the Boston and Maine, and the Central Vermont railroads, that would remain a dominant force in rail transportation well into the 20th century and beyond. The use of iron and later steel in railroad bridges by these, and many other lines, came about gradually, but the manner in which this change came about was rather contradictory. Just as covered bridges reigned supreme in road and highway bridge structures into the late 19th century, so, too, were rail companies well accustomed to building their bridges out of timber. Not only was the material inexpensive to procure and the technology to erect them well established, but wooden railroad bridges, both in covered and trestle form, over the years had proved to be solid and sound structures, a fact well evidenced by the survival of a number of examples of these types of bridges down to the present day. When iron was first introduced into railroad bridges in the 1840s with the introduction of the Howe truss, it was done so in conjunction with wooden materials, and this type of arrangement would continue for some time. Indeed, the still surviving Sulphite Bridge (nicknamed the "Upside Down" bridge), was built in Tilton, New Hampshire, by the Boston and Maine Railroad in 1896. However, as cautious, and, perhaps, cost conscious, as railroad officials may have been over the years, the introduction of heavier types of locomotives and rolling stock effectively ensured the gradual switch to iron and steel bridges as the standard for all but the smallest crossings on New England's busiest lines. Where small branch lines remained in use without heavy traffic some covered railroad bridges survived, but elsewhere they soon became a thing of the past.

Of the earliest iron railroad bridges built in northern New England, these were often two non-truss types, plate girder and trestle bridges. The first to see widespread use were girder bridges, either the deck-plate girder, which has no sides, or the through plate girder type, which has built up web plates riveted to the main support girder by angled components. These types of girder bridges were first developed in England in the 1840s and were introduced in America by the 1850s. Maine bridge engineer/historian, Llewellyn Edwards, identifies one company in his native state, the Portland Company of Portland, Maine, which fabricated a few of these bridges for the Atlantic and St. Lawrence Railroad (later the Grand Trunk Railway) about 1850. The Portland Company was best known for its manufacturing of locomotives, but is known to have fabricated the three "boiler plate girder" (Edwards, pg. 125) bridges erected at Portland and Gilead, Maine, and one over the Connecticut River at North Stratford, New Hampshire, based on the same design for bridges the railroad constructed in Quebec. These girder bridges, while outside the immediate scope of this work, are important and have been mentioned for two simple reasons; firstly, their successful use fostered the transition from wood to metal railroad bridge building, and, secondly, they proved to "possess greater durability than any other form of riveted bridge construction" (Edwards, pg. 126). Indeed, plate girder railroad bridges remain in extensive use on the railroads in the region to this day; over three dozen can be found in the incomplete records for New Hampshire, while about 90 have been documented in Vermont. A list of current railroad bridges for the state of Maine has been difficult to find, but anecdotal information and explorations suggests that the state also has a large number of plate girder bridges remaining, perhaps surpassing the number found in Vermont.

Trestle type bridges are another category of bridges that helped railroads span the gap, so to speak, from the era of all wooden crossings to that of iron and steel. A true trestle crossing may best be described as one in which the railway is supported, often across a gorge, wide valley, or roadway, rather than a river, on a series of supports called bents. These bents have four legs that are divergent from each other and are strengthened with cross members and bracing, much like a sawhorse, and are typically spaced close together. Wooden railroad trestles were a quite common feature on railroads throughout America, and are still in common use today in New England. Because the trestle form was so common on railroads, the term "trestle" is often erroneously used to describe any railroad bridge. An example of this is the Cuttingsville Trestle in Wallingford, Vermont, which is really a combination deck truss bridge. The first trestle bridges in the region made of iron probably came into use in the 1870s in northern New England, often in spectacular settings. The most famous example is the picturesque Frankenstein Trestle, located at Crawford Notch in the White Mountains of New Hampshire. Constructed by the Portland and Ogdensburg Railroad Company 1875, this 85-foot high, 520-feet long bridge was originally supposed to have been constructed of wood. But folklore has it that the ship transporting the pine timbers from the south was lost in a storm and the railroad subsequently turned to the Niagara Bridge Company to get the job done, utilizing wrought iron bents instead. The Frankenstein Trestle remains in use today, though it was rebuilt in 1892 with steel bents and strengthened for heavier railroad equipment in 1930 and 1950, and is one of the wonders of the White Mountains. Other examples of imposing trestle bridges are the Gulf Stream Trestle on the Somerset Railroad in Bingham, Maine, measuring some 115 feet high and 600 feet long, and the High Bridge in Claremont, New Hampshire. Trestle bridges were also used in other, more down to earth, settings for local electric streetcar systems throughout the region. One of the most interesting was the Letter "S" Trestle in Scarboro, Maine, on a line leading to Old Orchard Beach, a popular tourist destination.

In regards to iron and steel truss railroad bridges, this more advanced form would come to the region in the 1860s. The earliest truss types utilized by the railroads were Whipple or lattice truss, though by the mid–1890s the Warren, Pennsylvania Petite and Baltimore truss types were fast becoming the most commonly used forms. While the documentation of the gradual switch from wood to iron and then steel bridges by railroads in northern New England is incomplete, enough information exists for each state to give a general timeline.

The development of railroad bridges in the state of Maine is best illustrated by its most powerful railroad company, the Maine Central Railroad (MCR). The annual reports for the MCR, whose operations extended throughout Maine and into northern New Hampshire, clearly document the switchover from wood to iron and finally to steel. While it is unknown when the MCR built its first iron bridge, one of the earliest known examples was that over the Cathance River at Topsham, likely a Whipple truss structure built in the late 1850s to the mid–1860s. It was replaced with a new iron bridge, built by the Keystone Bridge Company of Pittsburgh, in 1878 because the earlier structure was considered too lightly built. Another notable early iron bridge was that built across the Kennebec River in Augusta in 1870, considered a beautiful and graceful structure by many. From 1870 to 1879 the MCR built 63 bridges in all; 14 of them (22 percent) were made of iron. How many of these may have been plate-girder crossings is unknown, but several, those built in 1877 at Falmouth (Presumpscot Falls) and Oakland (Rice Stream), were riveted lattice truss designs. From 1880 to 1889, the MCR built over 70 bridges, of which 44 (about 63 percent) were fabricated from iron. A mixture of wooden and metal bridges were built in the years up to 1887, but in that year 15 bridges were built, all of which were iron. Many of these early iron bridges were through truss structures that were later, ca. 1906–1912, replaced with steel plate girder or steel truss bridges. The change from iron to steel bridges on the MCR came about fairly rapidly; in 1891 and 1892 the company noted the first class iron bridges built at Mattawamkeag (3 spans, 476 feet total) over the Penobscot River, Bancroft (2 spans, 243 feet total) over the Mattawamkeag River and ten other locations ranging in size from 32 feet to 149 feet. The following year, in 1893, the MCR built its first steel bridge, a four span, pin connected bridge across the Sheepscott River at Wiscasset measuring 612 feet long. The next year, steel bridge building was in full swing on the MCR with the construction of the heavy steel, four-span bridges over the Kennebec River (614 feet) and Ticonic Falls (634 feet) in Waterville, replacing the iron bridges previously built there in 1874. By 1896, the age of the wooden bridge was largely at an end on the MCR, state records showing that the line had 150 iron or steel bridges on its routes.

The Gulf Stream Trestle on the old Somerset Railroad in Bingham, Maine, was an impressive structure measuring 115 feet high and 600 feet long. The span was built by the Boston Bridge Works in 1904 and served the railroad and its successor, the Maine Central, until it was abandoned in 1937. The bridge subsequently served as a roadway for hikers, hunters, and woodsmen until it went out in 1976.

While the Maine Central Railroad is one of the best documented in terms of its bridge building, it must be remembered that there were many other railroads active in the state. In looking at statistics for the year 1896, we find that the Boston and Maine, oper-

The Pittsburgh Bridge Company featured the Sandy River Bridge in Phillips, Maine, in one of their advertisements during the 1890s.

ating in the southern part of the state in York and Cumberland counties, had 32 iron bridges, though a 1914 valuation survey recorded no truss or through bridges in Maine for the B & M, indicating that many of its bridges were of the plate girder type. In the northern part of the state, the Bangor and Aroostook Railroad was an important concern and built some impressive bridges on its line, owing 55 iron and 33 wooden bridges. This railroad is active today, operating as the Montreal, Maine, and Atlantic Railway. Finally, in the central western part of Maine there was the unique two-foot gauge line of the Sandy River and Rangeley Lakes Railroad, which operated from 1879 to 1935. This profitable line, taking advantage of both the lumber business and a booming tourist trade, built two major bridges across the Sandy River and a tributary stream in Phillips and Strong by the 1890s. Other lines with a substantial amount of iron bridges by 1896 include the Portland and Rumford Railroad, with sixteen bridges, none exceeding 80 feet, and the Grand Trunk Railroad, with 67 such bridges measuring from twelve to 160 feet in length.

New Hampshire, too, was well on the way to building iron bridges by the 1870s. By 1884, New Hampshire had at least 35 iron railroad bridges, including two new truss bridges on the Suncook Valley Railroad, two 1883-built iron spans on the Boston and Maine at Exeter (60 feet long) and Dover (117 feet long), ten on the Concord Railroad, and 20 on the Atlantic and St. Lawrence Railroad in the northern part of the state measuring 20 plus feet in length, a number of which were probably plate girder bridges. From these accounts it is clear that New Hampshire, too, by the 1880s, was leaving the era of wooden bridge building behind. Indeed, the railroad commissioners for the state summed up conditions quite succinctly in 1884 when mentioning the Manchester and Keene Railroad; "Several iron bridges have taken the place of the discredited wood bridges and trestles" (*40th Annual Report*, pg. 31).

This transition would continue through the 1890s and the first decade of the 20th century. In the 1890s the Boston Bridge Works company built a number of riveted lattice truss bridges throughout the state, including the Fabyan Station bridge (1892) in the White Mountains, a similar structure at East Rochester, and probably the Ashuelot Bridge in the Westport section of Swanzey, while in 1909 they built a large, 3-span steel bridge over the Merrimack River at Nashua for the Boston and Maine, a bridge that is still in use today. Likewise, the bridges on the Maine Central Railroad line through the White Mountains saw significant upgrades in the early 1900s. The Second, Third, and Fourth Iron Bridges (as they are still called today) across the Saco and Sawyers rivers, all measuring between 153 and 164 feet long, were built in 1906 by the Pennsylvania Steel Company of Steelton, Pennsylvania. All these bridges are still in use today on the Conway Scenic Railroad. Another notable bridge on this line was also rebuilt in 1905, the Willey Brook Bridge in Harts Location.

The state of Vermont, like Maine and New Hampshire, saw a similar transition from wood to iron and soon thereafter steel railroad bridges in the 1880s. However, unlike the other two states, Vermont's efforts were spurred by several dramatic bridge disasters within a short span of time. In 1886 the Brattleboro and Whitehall Railroad's bridge at Three Bridges Crossing over the West River collapsed, killing two people and severely injuring five others when a locomotive and seven freight cars dropped 40 feet into the river below. This incident was followed six months later by the loss of the White River Bridge in Hartford on the Central Vermont line. This horrific disaster occurred in the early morning hours of February 5, when the Montreal Express passenger train derailed and then caught the "slight" (*New York Times*, Feb. 6, 1887) span on fire, causing both the bridge and four passenger cars to fall into the ice below, killing more than 40 people, many burned to death in the wreckage. Both bridges were rebuilt in iron by the Vermont Construction Company (VCC), utilizing a riveted lattice truss design. The new White River Bridge was an impressive structure, consisting of four spans and measuring 650 feet long, utilizing the old stone piers. The VCC proved the strength of the new bridge upon completion by loading it with twelve locomotives with a combined weight of 854 tons. Even more important, the possibility of a future fiery tragedy was now greatly reduced.

However, despite these dramatic events, Vermont, with its many short-line railroads, was perhaps a bit slower than its neighbors in leaving wood construction behind. The Vermont Public Service Commission's report on railroads in the state from 1910 to 1912 is quite interesting in highlighting bridge building activities. With the exception of the Boston and Maine (nine new steel bridges), the Central Vermont (15 new steel bridges), and the Grand Trunk, whose bridges were already considered above approach by the state's railroad commissioners, the state's smaller lines still had a fair amount of wooden bridges in service. Some, like those on the St. Johnsbury and Lake Champlain line, were well constructed and maintained. Other lines, however, struggled with their wooden bridges; the Montpelier and Wells River Railroad had mostly wooden bridges on its 45 miles of track that had been in service since the railroad began in 1873. The state commission admonished them that the bridges "are not adequate to the needs of modern traffic. They should be at an early date, replaced with modern steel bridges" (Vermont, *13th Biennial Report, 1910–1912*, pg. 366). Similarly, the diminutive White River Railroad, which ran nineteen miles from Rochester to Bethel, had as many as eight wooden Howe truss spans in 1910, though at least one was replaced with a riveted through truss bridge by 1912. The White River Railroad, known locally as the "Peavine" railroad because its tracks spread and curved hither and yon like a vine, had a colorful and troubled history in its struggle to survive, and one of its lattice truss bridges still stands today in Bethel. Finally, the Woodstock Railway was a small line that made a spectacular investment in 1911, replacing a wooden bridge across Quechee Gorge, near Deweys Mills, with a steel deck bridge designed by engineer John Storrs at a cost of $26,000.

Though largely outside the public view, the old iron and steel railroad bridges that remain in northern New England today represent an important and diverse group of spans; the oldest

Top: This impressive seven-span Pratt deck truss bridge over the Aroostook River in Ashland, Maine, was built by the Bangor and Aroostook Railroad in 1902 (courtesy of Maine Historic Preservation Commission). *Bottom:* This view shows an early iron bridge of the double-intersection Warren truss type across Otter Creek in Middlebury, Vermont.

in the region is the 1890-built St. John Street underpass bridge in Portland, Maine. This two lane, three-truss bridge was built by the Boston Bridge Works company for the Maine Central Railroad. This bridge is followed closely in time by the 1891-built bridge over the Passumpsic River in Barnet, Vermont, and the 1892-built Fabyan Station Bridge, a double-intersection Warren truss structure, over the Ammonoosuc River in Carroll, New Hampshire, also built by Boston Bridge Works.

While Maine has two and New Hampshire just one railroad truss bridge built prior to 1900, Vermont has ten. In addition to claiming the oldest metal truss railroad bridge in the state, Barnet also has an 1895-built pin and eye-bar connected truss bridge, a rare survivor. Other Vermont towns notable for their pre–1900 railroad bridges are Rockingham, which has two over the Williams River, and Middlebury, with two over Otter Creek.

In addition to these bridges, northern New England also has a number of other interesting metal truss railroad bridges still standing. In Maine there is the 1893-built Sheepscot River Bridge that consists of four spans, including one Pennsylvania truss and one draw span, that

The Iron Bridge in Springfield, Vermont, was an extremely tall Warren pony truss structure designed to carry the Springfield Electric Railway's tracks over the Black River. This unusual view shows the angled support beams that are attached to the bridge's sidewalk support beams.

measures 612 feet long, while the entire crossing, with causeways, measures about 1,000 feet in length, connecting the towns of Wiscasset and Edgecombe. This bridge, formerly owned by the Maine Central Railroad, is now owned by the state and lies on tracks operated by the Maine Eastern Railroad. Also in Maine there is a 1902-built deck truss bridge in Ashland on the former Bangor and Aroostook Railroad line, a seven span bridge that includes four pin connected steel Pratt trusses, spanning 777 feet over Sheridan Road.

In New Hampshire there is a 1903, pin connected Baltimore truss bridge in Haverhill across the Connecticut River, which once served both highway and rail traffic but is now abandoned; a 1905-built bridge over the Sugar River in Claremont that was owned by the Sullivan Machine Company, also abandoned; the Ashland-Bridgewater bridge over the Pemigewasset River, built in 1903 and designed by John Storrs; and the impressive 1905-built Willey Brook Bridge in the White Mountains at Harts Location. Finally, in Vermont, there is the impressive 1895-built Cuttingsville Trestle (not really a trestle, as previously discussed) over the Mill River in Wallingford, a combination deck truss and plate girder crossing measuring 412 feet long, the longest railroad bridge in the state; the impressive deck truss bridges at Ludlow over Andover Street, built in 1895 and measuring 286 feet in length; and Rockingham, built in 1894 and coming in at 191 feet long over Williams River; the 1903 built Warren truss bridge over the Winooski River, measuring 256 feet in two spans; as well as a smaller Warren truss bridge over the same river, both in Montpelier, built in 1902 and measuring 147 feet long with now rare surviving pin and eye-bar connections.

In closing the discussion on railroad bridges in northern New England, one cannot help but note the several heritage railroads located in the region. While many railroad bridges are off the beaten path and are not always easily accessible, a few groups of them can be viewed up close and personal by taking a ride on these historic railroads, New Hampshire has three such railroads: the Conway Scenic Railroad, which operates in the heart of the White Mountains; the Hobo Railroad, which operates out of Lincoln; and the affiliated Winnepesaukee Scenic Railroad operating between Meredith and Lincoln. In Vermont, there is the well established Green Mountain Railroad, with excursions offered on the Lake Champlain Flyer in the western part of the state, as well as on the Green Mountain Flyer between Chester and Bellows Falls, and the White River Flyer from White River Junction to Thetford. Finally, Maine has the Maine Eastern Railroad, operating from Brunswick to Rockland on the Maine coast, crossing some 30 bridges along the way, including several of the most impressive truss bridges in the entire region.

Whatever their location, and whatever type of bridge they are, the appeal of these old iron and steel railroad bridges today is simply yet eloquently summed up by the words of Kathryn Nelson of the Nashua River Watershed Association (NRWA). In describing the old Boston and Maine railroad bridge in Nashua, New Hampshire, over the Merrimack River, she comments that "the structure beckons from a time when railroads were the industrial force in the northeast — sturdy angular steel carrying trains and people over rushing water" ("Nashua River Recreation," NRWA, pg. 8).

8

The Death of a Bridge

Bridges stand and bridges fall; it's just a fact of life. The causes for the demise of any bridge are quite simple, no matter whether they be constructed of iron, steel, concrete, or wood. They can fall due to structural failure, the result of maintenance neglect and aging, accidents on the bridge itself, the forces of mother nature, or any combination of these factors, and they can be deliberately taken down for a variety of reasons already mentioned. Occurrences of strict structural failure due to poor design alone, though hinted at by some engineers as a possibility, is never known to have happened in the region. However structural failure due to age, poor maintenance, and exceeding load capacity has happened from time to time. Two interesting examples in New Hampshire show how such catastrophic events, widely spaced in time, happened to bridges large and small. In 1926 the small Grist Mill Bridge over the Mad River in Campton collapsed when a Lombard tractor with a load of logs was passing over, resulting in the death of the driver. This collapse was almost certainly due to the heavy weight of the tractor and its load, way too much for a bridge built in the late 1890s before the advent of modern transportation. On the other hand, there was the total collapse of a large 2-span bridge in Milan; this bridge had served for 75 years, being built by the Phoenix Bridge Company of Pennsylvania in 1889 for $9,500 and utilizing their patented iron column construction. The bridge fell into the Androscoggin River on March 20, 1966, while one car was passing over, without loss of life, likely due to a combination of age and a lack of maintenance. Other instances of these types of bridge loss have certainly occurred in many northern New England towns over the years, while many more bridges in danger of such occurrences have been closed. In some cases, bridges are rehabilitated and put back into service, but in many other instances these aged bridges are simply too far gone, and it is often more economically feasible to replace them altogether. Sometimes this process happens over the course of many years, but for more important spans it happens on a greatly accelerated basis. Vermont's Lake Champlain Bridge was closed almost without warning in October 2009 after an inspection revealed severe defects in its pier supports. While travelers scrambled to find new routes and the states of Vermont and New York hurried to provide additional ferry service, state engineers were hard at work to find a solution. With the old bridge unfit for travel, it was quickly determined that a new bridge was the only real solution, and in December 2009, less than three months after it closed, the Lake Champlain Bridge was demolished and work for the new bridge was underway. Still under construction as of this writing, the new bridge is scheduled for completion in October 2011. Likewise, the Memorial Bridge between Portsmouth, New Hampshire, and Kittery, Maine, was quickly closed after a safety inspection in 2011 and is scheduled for demolition in 2012.

By far the greatest instigator in change when it comes to the field of bridge building and design are the forces of nature and the calamitous results of severe water and wind action. While the action of wind and water action in relation to the direct strength of a bridge structure, as well as their indirect erosive effects around the bridge site, seems obvious to the layman today, the scientific principles in this area during the first decades of iron and steel

This Warren pony truss bridge collapsed in Campton, New Hampshire, while under a heavy load (courtesy of Campton Historical Society).

bridge building was not yet fully understood. Thus it was that many early bridges were not built to withstand the forces of nature. One common mistake, such as is illustrated by the Steep Falls Bridge in Limington, Maine, was building a bridge too close to a river or stream and not taking into account either seasonal flood activities or planning for floods of epic proportions, today often referred to as "hundred year" flood events. The same is also true when it comes to the principles of hydraulic action and its effects on bridge piers and abutments. The ever present wear on these bridge supports by constantly moving water, termed the "scour" effect by bridge engineers, was understood to a degree by early builders; but later developments, sometimes through painful trial and error, proved that bridge engineers also had to take into account the changes wrought when the natural flow of a river was constricted by piers and abutments built close to or on the water's edge. However, while the everyday effects of the forces of nature on early iron and steel bridges posed a challenge, it was the flood events in northern New England that caused bridge problems on a yearly basis. One source of these flood problems was the annual spring flooding that caused localized problems as the winter snows melted and turned small rivers and streams into raging waters, perhaps exacerbated by heavy foresting activity in some regions that resulted in greater runoff than normal. In May 1923, for example, the state of Maine suffered the loss of many bridges on the Kennebec River during heavy floods, resulting in the subsequent building of seven new metal truss bridges. But, it was those "once in a life-time" events that would literally change the bridge landscape forever, the catastrophic floods of 1927 and 1936.

The Flood of 1927

This event occurred in early November after the region experienced extremely heavy rains, some five to ten inches in a 24-hour period. The previous month of October had already

This view shows the bridge wreckage that resulted from the Flood of 1927 in Center Rutland, Vermont. In the foreground is shown the highway bridge partially washed out, and behind it is the Delaware and Hudson railroad bridge, what appears to be a double-intersection Warren deck truss structure, being swept away.

been a wet one, with rainfall totaling about fifty percent above the norm in Vermont. So when the heavy rains of November 2, 3, and 4 came, the ground was simply too saturated to absorb any more water, resulting in a flood of epic proportions. Vermont was particularly hard hit, with New Hampshire suffering lesser damage. The most affected rivers were the Winooski and White rivers, and Otter Creek in Vermont and the Connecticut, Merrimack, Ammonoosuc, and Androscoggin rivers in New Hampshire, as well as the many tributaries of these rivers, quickly rising to previously unheard of levels above flood stage. Vermont alone suffered 84 killed in the floods, 55 in the Winooski valley alone, with 1,258 bridges of all kinds damaged or destroyed to the tune of 4.5 million dollars, not including the damage to the state's railroad infrastructure. The Central Vermont Railroad alone lost numerous bridges and destroyed tracks that brought rail traffic to a standstill on its line for about three months, with losses totaling nearly three million dollars. Overall, the damage to Vermont railroads and electric railways totaled nearly four million dollars. Vermont in particular responded aggressively to the Flood of 1927 by initiating an ambitious building program to standardize bridge building in the state. As a result of the flood damage, new truss bridges, most of them Warren pony, Pratt, or Parker types (for short, intermediate, and longer crossings, respectively), were subsequently built, or rebuilt, in the towns of Berlin, Richmond, Royalton, Moretown, Sharon, Sheldon, Waterbury, Enosburg, Hartford, Montpelier, Richford, Barnet, Highgate, Johnson, Morristown, Newfane, Stockbridge, Warren, Rutland, Rockingham, Wolcott, Brandon, Ludlow, and Bridgewater during the years 1928–1929. These new bridges often replaced the more lightly built riveted lattice or lenticular truss bridges from the early days of iron bridge building that did not survive the floods and utilized sections of modern rolled I-beams rather than built-up members.

New Hampshire, which suffered damage to a much lesser extent than Vermont, just over

$2.7 million to bridges and highways combined, nonetheless replaced a number of bridges in the north country at hard hit Littleton and Bethlehem (two bridges each), as well as others at Twin Mountain, Jefferson, Dalton, Shelburne, Gorham, Bristol, Bath, Sugar Hill. On the Connecticut River, the state highway department built the 352-foot long Pennsylvania truss bridge connecting Piermont, New Hampshire, and Bradford, Vermont, the most impressive of the flood replacement bridges, as well as Parker truss span at Stewartstown. Like her neighbor, New Hampshire's bridge rebuilding program adopted the standardized designs and specifications of the federal Bureau of Public Roads.

Finally, the state of Maine suffered relatively minor damage in the Flood of 1927, mostly confined to the western part of the state in the Androscoggin River valley, and, like New Hampshire, suffered no loss of life. The most damage in Maine occurred to the railroad infrastructure, and even this was minor in nature. The Bangor and Aroostook Railroad reported $4,000 worth of damage, while the Maine Central reported $200,000 worth of damage, some of which was likely incurred on its Mountain Division branch in northern New Hampshire. As bad as the Flood of 1927 was, another flood of a lifetime would hit the region with even more devastating results in less than ten years.

The Flood of 1936

The Flood of 1936 was due to a combination of unusual weather events; a severe winter in 1935–1936 left a heavy blanket of snow still covering northern New England into early spring when, from March 11 to March 13, a warm front moved in bringing a deadly combination of snow-melting temperatures and heavy rains. Stalled in the region for several days, the front brought large parts of New Hampshire's White Mountains and northern Maine anywhere from 5 to 7 inches of rain. At first, just the smaller river valleys, such as the Pemigewasset and Saco, flooded; but by March 18, things worsened when the larger rivers broke free of their ice. Not only did they quickly rise above flood stage, but the massive amounts of ice remaining moved swiftly downstream wreaking havoc along the way. Many bridges were destroyed due to the combined force of the rushing water, ice, and debris slamming into them. Some bridges were even damaged or destroyed by neighboring bridges that were swept away and careened downstream. These bad conditions were further exacerbated by another large period of rainfall, up to ten inches in the White Mountains, in a period of just two days from March 18 to March 19, resulting in continued severe flooding and ice movement. By the time the flood was over, new records were set all across the region that, in many cases, have yet to be broken to this day. Hookset and Manchester, New Hampshire, were particularly hard hit on the Merrimack River, with the former town recording some 18 feet of water in its downtown area, while Manchester saw one of its most notable bridges, and one of the largest lenticular truss spans in New England, the two-deck McGregor Bridge, destroyed in the flood.

Likewise, the Connecticut River valley also suffered greatly, though few metal truss bridges were lost; several wooden spans were destroyed, but the bridges built after the Flood of 1927 were engineered well and survived the flooding. Vermont towns near the Connecticut River suffered as well, particularly at Vernon where a dam burst, but elsewhere in the state were generally spared the damage suffered by Maine and New Hampshire. In Maine, it was the Saco, Androscoggin, and Kennebec river valleys that were the most impacted; Brownfield, Limington, and Buxton on the Saco all suffered bridges losses. On the Androscoggin and its tributaries, heavy damage was incurred at Rumford, Jay, the twin cities of Auburn and Lewiston, Turner, and Durham-Lisbon, while on the Kennebec bridges were damaged or destroyed at such places as Augusta, Waterville, Richmond, Dresden, and Detroit, to name just a few. Unlike on the Connecticut River in New Hampshire where more modern bridges stood, most

FLOOD SCENE. WRACK AND RUIN AT HOOKSET, N. H.

TOPSHAM, ME., BRIDGE ON KENNEBEC RIVER GOES OUT

Top: The flood of 1936 caused great damage in New Hampshire's Merrimack River Valley. This view shows the damage in the village of Hookset and the washed away southern span of Lilac Bridge at bottom. *Bottom:* This railroad bridge over the Kennebec River in Topsham, Maine, was one of many Maine bridges to fall victim to the Flood of 1936.

of the bridges lost in Maine during the Flood of 1936 were either older wooden bridges, or older metal bridges that were more lightly built. One notable exception was the loss of three spans of the Kennebec Bridge at Richmond, a 1931-built six span swing bridge. In all, Maine had 150 bridges destroyed or damaged during the flood, 17 of them major crossings on large rivers. Indeed, of all New England states, Maine was probably the hardest hit of all.

Once again, just as happened after the Flood of 1927, the rebuilding process began anew, and where new steel truss bridges were built they were generally of the tried and true Warren, Pratt, and Parker types. It is interesting to note that in the current inventory of historic metal bridges built prior to 1940 in northern New England, Vermont has only three post-flood bridges from this era, reflecting the lessened impact of the Flood of 1936, while New Hampshire has ten, nearly equal to the amount of bridges built after the Flood of 1927. Maine's bridge woes during this second "hundred-year" flood in less than ten years time are well illustrated by the fact that it has twenty-one post–Flood of 1936 bridges, just under half of its entire remaining pre–1940 metal truss bridge inventory.

Because this flood occurred at the height of the Depression Era, the federal role in rebuilding northern New England bridges, as well as those in other states affected by the floods from Pennsylvania to Massachusetts, was greatly expanded in comparison to the Flood of 1927. The Bureau of Public Roads (BPR) offered design and construction oversight in coordination with state highway agencies, with the Public Works Administration (PWA) providing the financing. Interestingly, for the many smaller bridges destroyed in the floods, the Works Progress Administration (WPA) provided local labor to get bridges rebuilt, with town and county agencies handling the planning and finance. The alphabet agencies of President Roosevelt's New Deal administration, indeed, were vital in restoring New England roads and bridges.

While the devastating floods of the 1920s and 1930s brought great grief to the area, the overall end result was a better engineered and more modern transportation infrastructure in northern New England, a phoenix rising up from the ashes. The day of the wooden bridge was now long past, while that of the first generation of iron bridges, or what was left of them, had been literally swept away. The fact that so many of these immediate post-flood bridges still remain is a testament to how well engineered they were, and how important they continue to be to this day.

However, the forces of Mother Nature must always be reckoned with no matter how advanced bridge technology has become. This was evident again during Tropical Storm Irene, which struck the east coast on August 28, 2011, and left a trail of destruction. While Maine and New Hampshire were largely spared, Vermont suffered severe flooding, the likes of which had not been seen since 1927. Roads and bridges were washed out in many areas, resulting in many communities requiring supplies that had to be airlifted by helicopter. Towns such as Brattleboro, Wilmington, and Montpelier were quickly inundated and, as of this writing, the full impact of the devastation caused by the storm, though not yet fully tallied, will cost the state billions of dollars in repairs to roads and bridges.

9

Preservation Efforts

While the metal truss and other unique iron and steel bridges of northern New England have served their towns and cities very well for many years, many are now facing the biggest challenge of all; can they be preserved in a meaningful way within the modern transportation infrastructure? Before taking a look at these challenges, as well as some of the successes and failures achieved in the preservation arena, let's return to where this book began; the importance of the bridge in modern American culture. During the course of writing this book, friends and family were quite curious by my choice of the topic of metal truss bridges." Who cares about those rusty old bridges," one acquaintance of mine stated rather bluntly. Who indeed? I began to wonder if he might be right, until I started to pay attention to some of the television commercials I saw at various times — both day and night — and, lo and behold, what did I see? Truss bridges everywhere. Here was one in a backdrop for an American car company, there's another for an oil company ad, and yet another for a foreign car company. But, it didn't stop with these obvious choices for products related to the transportation industry. Here's another, this time for a nationally known woman's designer clothes brand (huh?), there was yet another for a famous brewery involving the sport of bungee jumping off — yes, you guessed it — a metal truss bridge, and one, too, for the U.S. Air Force! And, finally, to cite my last example (although there were many others, too numerous to list), there was the ad for a well-known skincare product comparing the human skin to a rusting truss bridge, by far the most clever of all the commercials I saw, and without a doubt the one most relevant today to the status of many truss bridges. Seeing all these "bridge" commercials out there in the visual and print media makes it exceedingly clear that metal truss bridges, whether we realize it or not, have themselves gained a unique place in our national psyche. Just as covered bridges were popular in print media for transportation related companies from the 1930s to 1950s, when Americans were taking to the road in ever increasing numbers; now it is the truss bridges, symbolic too of the open roads of America, that have taken their place. Can the grace, beauty, strength, even freedom, conveyed by the image of a truss bridge over a wide expanse of water or with a stunning sunset as a backdrop be matched by that of modern concrete and steel highway bridges in a similar setting? Of course, at the other end of the spectrum, the image of a rusting bridge is almost synonymous with that of urban blight in America. So, when we as Americans, whether it be local, state, or federal officials, or local citizen and historic groups, are making decisions about the fate of these "rusty old bridges," we must ever keep in mind that we are not just preserving a bit of local history, but are also collectively adding to the preservation of our national transportation heritage. While not all of these old bridges can, or even should, be saved, each time a decision *is* made in favor of preservation, another link with our open road past remains for future generations to enjoy.

The challenges in regards to saving metal truss bridges and similar structures are not new; engineers have been dealing with them for years. And it is only relatively recently, with the attention paid to America's aging infrastructure and the much publicized "red-list" of bridges in every state that are in dire need of repair, that there has been widespread public interest. The greater part of this interest has been generated by specific occurrences of bridge failures, none more stunning than the collapse of the I-35W Mississippi River Bridge in Minneapolis, Minnesota, a 13-span deck arch truss and girder bridge completed in 1967. Thirteen people were killed in the collapse on August 1, 2007, during the height of the evening rush hour, with nearly 150 people injured. And the event, at least for a time, shook the confidence of Americans nationwide. What other bridge disasters were out there waiting to happen? The fact of the matter is that every state, not just those in northern New England, has a backlog of bridges that need attention, and lacks the funds to get them fixed. This is the constant problem for America's transportation infrastructure as a whole. The one term that is most often used when referring to metal truss bridges is "functionally obsolete." Whether a bridge in question is structurally sound or not, most historic truss bridges are considered obsolete because they do not conform to modern bridge standards in regards to the loads they can carry, their height and width restrictions, as well as their sometimes dangerous approach situations. These points are well illustrated by two New Hampshire bridges; the Pingree Bridge in West Salisbury and the Sewalls Falls Bridge in Concord. The Pingree Bridge, built in 1893 to replace a covered bridge, was a relatively small crossing. Though in a rural area, and rehabilitated in 2006, the bridge continued to pose problems for the town and local residents. Very narrow in width, this one-lane bridge was simply too small to provide local access to delivery trucks, fuel-oil trucks, as well as larger fire and rescue vehicles. Despite the historic nature of Pingree Bridge, it was eventually decided, in 2009, that the bridge had to be replaced. This same debate is still ongoing for the Sewalls Falls Bridge in Concord, a poster-child of sorts for functionally obsolete structures. Built in 1915, this Pratt truss 2-span structure can only accommodate one lane of traffic, with drivers using the honor system, so to speak, in yielding to oncoming traffic. Because the bridge is located on a busy road serving a growing population and a mixture of residential areas and businesses, a solution to the problems regarding this historic yet outdated span will soon have to be found. Among those that use Sewalls Falls Bridge, opinions are divided. Some would like to see the rusty bridge (its poor appearance yet another factor in how people view this bridge) replaced with a modern span that can accommodate two-way traffic; while others, despite the inconvenience, appreciate the historic nature of the structure and would like to see the bridge preserved. The Sewalls Falls Bridge might have been replaced years ago, were it not for the fact that it is the sole surviving bridge designed by John Storrs remaining in his hometown.

Other constant problems regarding the preservation of historic truss and suspension bridges in northern New England are related to ongoing maintenance issues that are extremely expensive to solve. Most of these historic bridges have far exceeded their normal life expectancy, a testament to their designers and builders, but problematic for local and state highway departments today. Indeed, civil engineers a century ago were realistic about the structures they built, one bridge textbook stating that "the average life of iron or steel railroad bridges is probably not far from twenty years" (Merriman and Jacoby, *Roofs and Bridges*, Part III, pg. 15). Common maintenance issues include worn out pier supports, as was the case with Vermont's Lake Champlain Bridge and the Meadows Bridge in Shelburne, New Hampshire; or with worn wire cable supports in the case of Maine's Waldo-Hancock suspension bridge; all of which problems caused the bridges in question to be removed.

Painting these bridges, especially larger spans, is also problematic; the costs to strip historic bridges of their old paint are ofttimes astronomical because the paint is much more toxic than the paints of today and create both environmental and worker health concerns. This issue was a constant problem with the Memorial Bridge between Portsmouth, New Hampshire, and Kittery, Maine, over the years, but has also been a problem for smaller bridges as well. The workers on the recently renovated Rice Farm Road Bridge in Dummerston, Vermont, had to take proper precautions in this area during renovations in 2010, while the Watson Road Bridge in Dover, New Hampshire, sits on dry land, its fate in limbo for a number of years now because of the high costs involved in stripping the bridge prior to finding it a new home. Of course, because these bridges are expensive to repaint, and also due to a lack of timely funding, many are neglected to the point that they have become rust streaked and spotted. This often spells the death knell of a historic bridge (or any bridge for that matter); not only does rusting lead to structural degradation, but it also leads to public perception problems as to the worth of the structure. After all, it is difficult to garner public support to save a historic bridge that has so long been a blight on the local landscape that most citizens have a hard time seeing beyond the rust, and just want it replaced with something better looking.

Preservation Options

Today, there are a number of preservation options for metal truss bridges and other similar historic bridges. While prevailing conditions will dictate that not all pre–1940 truss bridges can be saved, it is important, at the very least, that representative examples of each truss type be saved, as well as those bridges that are more unusual in design. This approach is no different, in fact, than that which was taken by regional state highway departments in regards to covered bridges from the 1940s to the 1960s. New Hampshire and Vermont alone each once had well over 400 of these historic bridges, while Maine had considerably less (125 have been documented), but their numbers dwindled considerably from the 1890s to the 1940s as new iron and steel bridges were built to take their place. Interestingly, folks during this time period often considered some of these run-down covered bridges unwanted relics of the past and wanted them replaced with more modern structures. Unfortunately, the base amount with which today's engineers and preservationists have to work with among bridges to save is much smaller. Vermont has about 100 metal truss bridges remaining today, New Hampshire 73, and Maine about 50, exclusive of railroad bridges. We simply cannot afford to suffer the same percentage of losses as occurred with the region's covered bridges, or only a handful will be saved. So, what options are out there to keep these bridges going? Here are some of the most common options used in bridge preservation, presented in order of historic preference. In many cases, a combination of these methods, such as relocation and bypass options, have been used when preserving a bridge

Continued Highway Use

Where feasible, this is always the best plan when it comes to saving historic metal truss bridges. If it is kept within the system, and is properly maintained, it will continue to be of value and function as its builders originally intended. In many cases, bridge substructures, supports, and roadways have been, or will need to be, upgraded for today's heavy loads and safety standards. The Connecticut River bridges are a perfect example of how larger bridges can be rehabilitated to maintain their unique status in the local landscape. On a more localized level, there are many examples throughout northern New England of metal truss bridges being rehabilitated and preserved for continued use, among the most significant being the Ryefield Bridge in Harrison, Maine; the Rice Farm Road in Dummerston, the Paddock Road Bridge in Springfield, the Blacksmith Shop Bridge in Tunbridge, and the Granite Street Bridge in Montpelier, all in Vermont; and the Crane Hill Road Bridge in Sugar Hill and the Dow Avenue Bridge in Franconia, both in New Hampshire.

Relocation

Where a bridge cannot be saved in its original location, many times the trusses can be preserved and utilized in other locations. This unique option is attractive to many, but has its difficulties. The first of these is that the bridge is taken out of the original context in which it was built. However, removal is always preferable to permanent destruction, so the next question becomes at what location will the bridge be re-erected? Ideally, a relocated bridge will stay within the town in which it was originally built, but this need not always be the case. The state of Vermont and its Historic Bridge Program (HBP) has been masterful in planning reuse strategies for otherwise endangered bridges and keeping them in the public eye. This the HBP has done with bridges large and small. Their very first relocation project was a 1925-built Warren pony truss bridge in Hinesburg which originally stood on Turkey Lane over Lewis Creek, but now serves as a pedestrian bridge over a small creek on a path leading to the local post office. The Vermont HBP's most ambitious project to date has been the relocation of a 1902-built 2-span Pennsylvania truss bridge, measuring over 300 feet long, that once stood in West Milton. Once a new bridge was built in West Milton, the old bridge was no longer needed or wanted by the town. It was eventually removed and re-erected in Swanton in 2008 in the same location formerly occupied by a historic covered railroad bridge that had burned. One town did not want or need the historic bridge, but another did, thus resulting in a saved bridge and a preservation problem solved.

Relocation strategies in New Hampshire have been utilized to a much-lesser extent, and in Maine hardly at all. In New Hampshire, unwanted bridges, usually small pony-truss spans, have been sold to private individuals by local towns for the standard $1 dollar price so that it is taken off their hands, as occurred with the Pingree Bridge in Salisbury and the 1907-built Steele Road Bridge in Thornton, sold to a private snowmobile club for use on one of their pathways. This kind of option, while preferable to ultimate bridge loss, usually takes a bridge out of the public eye and may result in just delaying the loss of the bridge, as there is no oversight into future use and maintenance of the bridge by its private owners.

In Maine, as far I've been able to tell, the relocation option has been used very infrequently. The only example I've found is that of the old king-post truss Tannery Bridge, formerly located in St. Albans, which was removed and re-erected on dry land outside the DOT facility in Pittsfield. An incident that happened in Harmony, Maine, is indicative of some of the problems associated with selling old bridges to private individuals, and is probably not an isolated occurrence in New England, where towns have control of their local bridges. The Bailey Bridge (truss type unknown) in Harmony had been closed for over ten years when, in 1999, having been deemed a useless bridge, it was sold by a local selectman for $250 for use on a private narrow gauge railroad without putting the matter to a vote. Once the matter came to light, bickering among town officials ensued; and while the railroad owner offered to return the bridge, the bridge was later reported to have vanished and its ultimate fate is unknown. Finally, when discussing the option of relocation, the question of preserving and or storing the bridge trusses until they can later be re-erected is sometimes problematic. The state of Vermont has solved this problem by creating a bridge depot at Clarendon, where structures slated for relocation are properly stored until decisions about their future locations are made. However, in some cases such care is not always taken, in New Hampshire, two bridges that have been removed for possible relocation, the Meadows Bridge in Shelburne and the Watson Road Bridge in Dover, still sit intact on the river banks near where they formerly stood, awaiting their fate. The ultimate example of a displaced bridge, however, is that of the Murray Road Bridge in Bennington, Vermont, a Moseley iron arch bridge taken down in 1958, and, with an eye to its preservation, presented to the Bennington Museum in the hope that it might one day find a new home. Over fifty years later, the trusses of this bridge still lie in several pieces behind a museum storage building, its builders plate intact, still waiting for a home.

Bypass

For many years, starting with old covered bridges and stone bridges, the option of bypassing a bridge, by leaving it in its original place and building a bridge nearby as a replacement, has been a common preservation remedy. This has the favorable result in leaving the bridge in its original location. However, other factors must be taken into account when the bypass option is considered. First and foremost of these is the matter of who will take responsibility for the continued preservation of the bridge, and whether the necessary funds will be committed in the long-term for its upkeep. While the bridge will, perhaps, not have to be maintained to the same standards as a functioning highway bridge, it will still need painted periodically, its deck repaired or replaced from time to time, and attention given to its substructure. If the right uses for a bypassed bridge can be found, these funds will come much easier. By leaving the bridge open to pedestrians as a historical attraction and picnic spot, or as part of a hiking or biking recreation path, the bridge's public connection is maintained; and though its use has been altered, it remains a vital part of the community infrastructure, much like a public park or playground. This is the ultimately desired result when the bypass option is chosen. This is well exemplified by the preservation of Thunder Bridge in Chichester, New Hampshire, the Gambo Falls Bridge in Windham, Maine, and the Chesterfield Bridge, a 1936-built Connecticut River arch bridge between Chesterfield, New Hampshire, and Brattleboro, Vermont. However, there are situations where the bypass option is either not fully thought out, or becomes an unintended consequence of changing traffic patterns. An example of the former is the Medburyville Bridge, an 1896-built double intersection Warren truss bridge in Wilmington, Vermont. This bridge was the first in the state to be saved from demolition under new federal guidelines for bridge preservation. Unfortunately, no plans were made for made for the future use of the bridge; and today, while in good enough shape, the Medburyville Bridge just sits there. Closed to pedestrian traffic and with no deck, the bridge's purpose in the future is as of yet undefined. Finally, there are also several examples of bridges in the region that have been bypassed due to changing traffic pattern, usually those on roads between two towns. In this case, the road and the bridge are unwanted and, with neither town willing to keep up needed maintenance, the bypassed bridge soon falls into disrepair and, in effect, becomes an abandoned bridge. Examples of this type of situation are found most commonly in New Hampshire, including the Parker high truss bridge, built in 1907, over the Merrimack River between Boscawen and Canterbury; the 1893-built Thompson Crossing Bridge, a high Pratt truss bridge between Antrim and Bennington; the Cavender Road Bridge, a 1905-built pin-connected Pratt pony truss bridge between Greenfield and Hancock; and the 1910-built Jones Crossing Bridge in Milford, a high Pratt truss bridge.

Bridge Alterations

This option for "preserving" a bridge is one that is avoided by historians, but sometimes resorted to by local town officials, perhaps as a poor compromise between maintaining a bridge's historical nature and gaining a newer, and more modern bridge. In cases such as this, a bridge is converted or altered to a more modern type, such as one supported by steel I-beams or concrete, but the trusses are left in place as mere decorations, to retain the former look of a bridge, or, at best, to offer support, not to the roadway itself, but to its sidewalks. By altering and diminishing a bridge in this manner, the bridge truss, the very essence of a bridge, no longer functions as originally designed and serves instead as mere ornamentation. Historic bridge engineers dislike this option, though its impact may not be discernible to the layman. An analogy to this might be, say, preserving a Model-T Ford by lovingly restoring its body and interior to original specifications, but replacing its original engine with a modern V-6. All of us would agree that, while this "enhanced" Model-T would have more power and go faster, the original integrity of the car has been lost by such "improvements." The same applies to historic metal truss bridges. When restored properly their trusses can function with the original integrity with which they were designed. Examples of this type of altered bridge are found in Vermont on the Elm Street and Mill Street bridges in Woodstock, the Vine Street Bridge (a relocated span) in Northfield, and the Langdon and School Street bridges in Montpelier. Only one example of this has been found in New Hampshire, the Chef Road Bridge in Jackson, and none in Maine. The most bizarre example of an altered bridge, by far, is the Mill Street Bridge in Woodstock, Vermont. Here, the town decided to reinforce a ca. 1900-built Pennsylvania truss span with the addition of a modern steel arch bracing system designed by a nationally known bridge architect. The result of this unfortunate attempt is a strange hybrid bridge that looks like a giant spider or maybe a creature from outer space. Indeed, one Vermont bridge historian commented that this bridge was an example of how not to preserve a bridge, and the results speak for themselves!

Railroad Bridge Solutions

Because railroad bridges are a unique category all their own, the options for preserving them are somewhat more limited. However, because many of them have seen relatively light traffic, or none at all, over the last several decades, many are in fairly good shape. Seldom are railroad bridges, probably because of their size, considered for relocation, and the traditional bypass option is usually not in play. Instead, the best option for these bridges is either the continued activity on the rail lines they serve, whether it be freight or passenger traffic, or the restoration of such service. Excellent examples of such service is found in all three northern New England states in the form of the previously discussed local heritage railroads; local freight carriers, such as the Washington County and Clarendon and Pittsford railroad short lines in Vermont; as well as regional and national carriers, including Amtrak's Downeaster in New Hampshire and Maine and its Ethan Allen Express in Vermont, each of which crosses over at least one metal truss bridge somewhere along their routes. Failing active service, the best option for the preservation of these bridges, as well as the abandoned rail paths on which they stand, is the conversion of these abandoned lines to recreational rail trails, suitable for hiking, biking, and snowmobiling. This has been a popular option throughout the northeast, and has served to keep many former railroad bridges in service.

Funding Preservation

Whatever method of preservation is chosen, the most difficult problem of all in restoring these historic bridges, beyond functional obsolescence, costly maintenance, and public indifference, is finding the funds to pay for their upkeep and/or restoration. State and local budgets for the transportation network have been stretched to the limit now for years, and there simply is not enough money, even with federal funds, to fix all our bridges. Maine has 386 bridges on its "watch list" as of 2010, while New Hampshire has 146 bridges of all types on its "red list." In terms of national rankings in respect to the percentage of deficient bridges, northern New England scored worse than most other states and better than the worst states in the nation (Rhode Island and Massachusetts), but not by much; New Hampshire (31.35 percent —

The 1929-built Taylor Street Bridge in Montpelier languished for many years, but was restored in 2010 with funds from President Obama's stimulus program.

38th place) ranked best in all of New England, followed by Maine (34.23 percent — 42nd place), while Vermont (35.66 percent — 44th place) ranked third worst in New England, trailing Connecticut. While every year some bridges are repaired and removed from these state lists, others are added, with Vermont bridges leading the decline in northern New England. Heavily traveled highway and local bridges, including some metal truss bridges, are given priority for repair, while other bridges on the list have to wait their turn, continuing to deteriorate in the meantime. With the recent economic downturn nationwide, from 2008 to 2010, even regular federal highway funding has decreased. One potential bright spot for metal truss bridges was President Barack Obama's American Reinvestment and Recovery Act, signed into law in 2009 and popularly known as "the Stimulus." Among the many purposes of this act, including creating vital jobs for Americans during a period of economic woes not seen in this country since the Great Depression of the 1930s, was to provide funding to repair America's crumbling infrastructure. Of the $787 billion dollars in spending appropriations provided for in the bill, $27 billion was provided for highway and bridge construction and repair, and nearly $12 billion in mass transit and railroad-related projects, and it was the hope of many that historic truss bridges might gain a portion. To that end, James Garvin, State Architectural Historian of the New Hampshire Division of Historical Resources prepared a position paper entitled *Preservation and Reuse of Historic Bridges as Economic Stimulus*, which was supported by a number of regional preservationists and sent to the federal government. In part, Garvin states that "the present moment offers an opportunity for action.... Preservation of historic bridges is in keeping with longstanding public policy.... It is economically beneficial, inasmuch as rehabilitation, while usually less costly than new construction, is labor intensive and thus generates the need for many skilled jobs" (Garvin, pp. 3–4).

Whatever the hope the Stimulus might potentially have offered in terms of preserving metal truss bridges, however, it has not materialized. Indeed, just the opposite seems to have occurred; while a number of new bridges have been built nationwide as a result of the Stimulus, it has often been at the expense of older metal bridges. In northern New England, few metal truss bridges have actually been restored as a result of the Stimulus, including none thus far in New Hampshire. In Vermont, several bridge projects have been funded by the Stimulus, the Bridge Street Bridge in Richmond and the Taylor Street Bridge in Montpelier, as well as replacement decks on ten bridges on the Delaware and Hudson rail trail in Poultney, Pawlet, and Rupert, several of which may be truss bridges. In Maine, the historic Deer Isle-Sedgewick Bridge, a Steinman designed suspension bridge, received a $5 million grant to rehabilitate the bridge's substructure. One project failing to gain funds under the Stimulus was the vertical lift Memorial Bridge between Portsmouth, New Hampshire, and Kittery, Maine. Some $90 million had been requested for refurbishing the historic bridge, but it was denied. Instead, in October 2010 $20 million in funding was received under a TIGER grant to replace the aging bridge. The disbursements made under the Stimulus plan, in the end, did not benefit endangered metal truss bridges to the extent that was originally hoped for. Hope for the needed funds to restore historic iron and steel bridges in northern New England was revived yet again in September 2011 when President Obama proposed the American Jobs Act and asked Congress to pass the bill in quick order. Among its many provisions to help the economy is a provision for more funding to restore crumbling roads and bridges, thus creating thousands of construction jobs while helping to restore the nation's infrastructure. However, with the current political climate in Washington, D.C., it is unclear whether this bill has enough support, and needed bridge repairs still face an uncertain future.

In ending our discussion on preservation issues, it is important to note that one major problem for historic bridges in need of renovation hoping to gain Stimulus funds was one of the major requirements of the Stimulus itself, namely that a project be "shovel ready"; in other words already planned, developed, and ready to proceed once funding was received. One of the major reasons for a lack of preplanned rehabilitation projects is the fact that, except in Vermont, comprehensive historic bridge preservation programs in the region, as in many other states, is generally lacking. Thus it is that the desire by state transportation departments to replace, rather than preserve, aging bridges, especially those of the metal truss variety, often seems to be the driving force. New Hampshire and Maine have no formal historic bridge programs as a component of their state departments of transportation, though New Hampshire has proposed the formation of one. In Maine, the Historic Preservation Commission (MPC), which has eleven members, including a representative from the department of transportation, serves as a liaison with other state agencies where historical properties and matters are concerned. In New Hampshire, it is the Division of Historical Resources (DHR) within the Department of Cultural Resources that interacts with other state agencies, such as the Department of Transportation, to identify historic assets within the state and aid in formulating preservation plans where appropriate. While these agencies in Maine and New Hampshire have done great work in promoting metal truss bridge preservation and have extensively documented both bridges remaining in their inventory and those scheduled for replacement, the lack of a formal state program for preserving historic bridges has left them largely powerless and sometimes at odds with both local towns and cities, which maintain jurisdiction over bridges on all roads except federal and state highways, as well as with state departments of transportation. The lack of a formal historic bridge program does not mean that Maine and New Hampshire have not had some successes in the preservation arena. In New Hampshire, the efforts of the DHR were extremely valuable in the preservation of Thunder Bridge in Chichester, an impressive lenticular truss span over the Suncook River; while in Maine, the rehabilitation of the Gambo Falls Bridge at Windham and the Ryefield Bridge in Harrison, the only double intersection Warren truss bridge left in the state, was accomplished with the guidance of the MHC. These examples are just a few of the local success stories achieved in New Hampshire, and to a lesser extent in Maine, when it comes to preserving metal truss bridges in these states. However, it is the state of Vermont that leads the way. With the establishment of its Historic Bridge Program (HBP) as a division of its Agency of Transportation in 1998, it has not only become committed to saving specific historic metal truss bridges (as well as other historic bridges), but has a formal plan for doing so. Among the metal truss bridges saved and relocated through their adaptive use program are those, previously mentioned, in Swanton and Hinesburg, as well as others in Barre, Barton, Northfield, Thetford, Wallingford, West Rutland, and Westfield. In addition, the HBP has a number of bridges saved from destruction now placed in storage at Clarendon and awaiting re-erection elsewhere, including the old Howard Hill Bridge from Cavendish, the Pioneer Street Bridge from Montpelier, a lenticular bridge from East Bethel, and about eight others. Funding, as always, remains a challenge even with the establishment of the HBP in Vermont, but nonetheless a solid framework for preserving these bridges is now permanently in place. This is what is needed in every state, not just northern New England; the formal mandate to preserve metal truss bridges. This, combined with heightened public awareness, will ensure that these important structures survive for future generations.

PART II

Notable Bridge Histories

The following chapters detail some of the most historic and notable iron and steel bridges remaining in each state in northern New England today. In the state by state entries, in addition to listing bridge locations by town, I have also provided key statistics for each bridge listing details as to year of building, truss type, the feature crossed — be it river, stream, roadway, or railway — length, number of spans, and ease of access information for those visiting a given bridge in person. In regards to this last detail regarding bridge access, this is based on my own experience viewing and photographing these bridges up close and in person and my own judgment as to how well the bridge can be viewed from a variety of settings, not just on its deck as one crosses over by vehicle or on foot. Bridges are categorized by state, with an additional category for interstate bridges, which highlights the Connecticut River bridges between New Hampshire and Vermont (though technically these are under New Hampshire jurisdiction) and the Piscataqua River bridges between New Hampshire and Maine, which are owned jointly by an interstate bridge commission. Bridges slated for demolition are noted with an * after their name. While these bridges' days are numbered, their historical importance nonetheless warranted their inclusion.

Within each bridge history I have commented on what makes these structures notable, significant design features, and the builder where known. Though my decision about which bridges to highlight is partially subjective, an entry for every remaining bridge would result in some redundant histories and therefore the focus has been on solid examples for each truss category, as well as those that are generally recognized by historians as outstanding structures. If a bridge the reader is looking for does not have its own history, it may be found in the appendices at the end of the book which list all known remaining pre–1940 metal truss highway and pedestrian bridges for Maine, New Hampshire, and Vermont.

10

Vermont Bridges

1. Wilmington — The Medburyville Bridge; DATE BUILT: 1896; TRUSS TYPE: Warren Through, Double-Intersection; FEATURE CROSSED: Deerfield River; LENGTH: 96 feet; SPANS: 1; EASE OF ACCESS: Restricted

The Medburyville Bridge, Wilmington (courtesy Diane Chapman).

This bridge is one of just four multiple-intersection Warren truss bridges remaining in the state, most of which were built in the 1890s. Located in a pastoral country setting, the bridge served highway traffic until the 1990s, when it was bypassed and a new bridge was built to take its place. The preservation of this bridge marked the first time in Vermont that federal funds normally provided for in the demolition of old bridges was applied to bridge preservation. While the bridge is in reasonably good shape today, and a state historic marker is placed at the Medburyville Bridge, the bridge is closed to pedestrian traffic and has no deck. Located on the old Molly Stark Trail (Route 9) in the Medburyville section of Wilmington,

Harry Morse and his daughter Marilyn (age 5) are sawing wood while the diamond-shaped trusses of the Medburyville Bridge can be seen in the background (Photograph courtesy of Marilyn Morse Howard).

the bridge has long been a part of the local landscape and is fondly recalled in the childhood memories of those who grew up in its shadow. One such person was Viola Bishop Morse. Born in 1908, she set down some of her school days memories in the 1980s and recalled, "When I was little I used to walk to school with the older ones. Going across the bridge, we were always told not to run and to walk on the upper side because the wind blew so sometimes. I remember once they were putting new planks on and we had to walk the stringers ... one of the men took my hand and helped me across." (Morse manuscript, n.p.). Whether it was winter or summer, Viola's memories of the bridge have remained vivid all her life, even to this day. In the summer, during her first grade year, she and her friend, second-grader Henry Jacobs "went down to the river by the bridge ... and we started damming up the river." In the early spring the bridge was severely threatened by ice: "Sometimes when the river broke up, the teacher would let us go out and watch. Huge cakes would lift and float down and tear anything in front of it apart. They would fill the road ... sometimes 5–6 feet high" (*ibid.*). Later on, Viola Bishop would marry Harry F. Morse and live on the Morse Farm, located within sight of the Medburyville Bridge. Indeed, the structure always served as a backdrop to everyday life on this family farm, not only for Viola and Harry Morse, but for their children Marilyn and Charles as well.

---

2. Dummerston—The Rice Farm Road Bridge; DATE BUILT: 1892; TRUSS TYPE: Hilton Truss; FEATURE CROSSED: West River; LENGTH: 198 feet; SPANS: 1; EASE OF ACCESS: Easy

This recently (2010) renovated bridge is notable for its unusual truss type, age, and size.

The Rice Farm Road Bridge, Dummerston, Vermont. This view shows the bridge during its 2010 restoration.

Built by the Berlin Iron Bridge Company of East Berlin, Connecticut, the Rice Farm Road Bridge must have been considered by that prolific company as being among its greatest achievements as it was featured in their 1892 bridge catalog. This bridge is also the fourth oldest metal truss bridge in Vermont, and is just one of two among the oldest metal bridges remaining in use today in the state that has not been modified or relocated. Situated off busy Route 30, just north of the West Dummerston Bridge, the longest covered bridge in the state, the Rice Farm Road Bridge, too, is notable for its size. In fact, no metal truss bridge remaining in the state built prior to 1911 is longer, and this bridge is the longest Warren truss type thru bridge remaining in Vermont today. However, it is not its age or length that make this bridge truly distinctive, but its truss type. The patented Hilton truss was a variant of the reliable and standard multiple-intersection Warren truss that was developed for railroad use, and this is the sole bridge of its type remaining in the state today. How many other Hilton truss bridges were built in the state, or the region for that matter is unknown, but the number was likely small. The extra diagonals in this quadruple intersection riveted design, criticized for its excessive use of materials and the fact that its stress limits could not be accurately measured, offered rigidity to the bridge and a degree of impact resistance not found in regular highway bridges of the time. The history of this bridge clearly tells us why the Rice Farm Road Bridge was built so strong. It was built to serve the nearby Lyon Granite Company, which, with its modern technology, was quarrying ever-increasing sizes of granite blocks and needed a bridge strong enough to carry these loads. Interestingly, the abutments for this bridge are also made of granite, probably scraps from the granite company it served. It is the strength of this bridge, despite its narrow 14 foot width, that has made it a survivor, no doubt helping it to withstand the devastating Flood of 1927 when so many other early iron bridges were swept away.

---

3. Jamaica — Wardsboro Brook Bridge; DATE BUILT: 1936; TRUSS TYPE: Pratt through; FEATURE CROSSED: Wardsboro Brook; LENGTH: 124 feet; SPANS: 1; EASE OF ACCESS: Easy

The Wardsboro Bridge, Jamaica (courtesy of Craig Hanchey).

Located just off Vermont State Route 100, this abandoned bridge sits on what is now an abandoned road and is closed to traffic. The area in which the bridge is located, close to East Jamaica, was once the center of the town's activities, but became less populous as its settlement shifted to the west toward Manchester and the upper reaches of the West River. This bridge is one of those structures that, while not noteworthy in design, was once important in keeping the many small settlements within the town of Jamaica connected. Though small in numbers, the families that settled in Jamaica were widely spread out and at one time there were ten school districts in the town, each with its own schoolhouse. The bridge was also important to local industry, though this is not readily apparent today; a number of sawmills were once located near the site of the Wardsboro Brook Bridge. The future fate of this bridge is in doubt, though it is a possible candidate for relocation in Vermont's Historic Bridge Program.

---

4. Wallingford — The Cuttingsville Trestle; DATE BUILT: 1895; TRUSS TYPE: Combination Warren deck, plate girder, and arch; FEATURE CROSSED: Vermont Route 103 and Mill River; LENGTH: 412 feet; SPANS: 3; EASE OF ACCESS: Moderate

The Cuttingsville Trestle, Wallingford. This view shows its center truss span, with a deck plate girder span to the left. On the bridge are several diesel locomotives of the Green Mountain Railroad, as well as a locomotive being used under lease from the Santa Fe Railroad (courtesy of Ryan Parent).

This railroad bridge is notable for being the longest in the state overall, though non-combination railroad deck truss bridges that are longer than the truss portion of Cuttingsville Trestle may be found in New Haven and Ludlow. The bridge was once part of the famed Rutland Railroad, originally chartered in 1843, and is located on a branch line ending at Bellows Falls. However, when the Rutland Railroad went bankrupt in 1963, the state of Vermont gained control of much of its trackage. Today, the state-owned railroad system, Vermont Railway, owns the bridge, with its Green Mountain Railroad division operating over the old Rutland branch line. The Cuttingville Trestle continues to serve rail freight and passenger traffic and its location in East Wallingford is a popular viewing spot for train enthusiasts and photographers alike. As previously mentioned, though referred to as a "trestle," this bridge is in fact a combination deck truss bridge, utilizing several of the more popular railroad bridge forms, with the truss portion located high over the Mill River and the plate girder span over normally dry ground, while an arch bridge spans the highway.

---

5. Springfield — The Paddock Road Bridge; DATE BUILT: 1929; TRUSS TYPE: Baltimore; FEATURE CROSSED: Black River; LENGTH: 160 feet; SPANS: 1; EASE OF ACCESS: Easy

The Paddock Road Bridge is one of only two Baltimore truss highway bridges in Vermont.

Side view of the Paddock Road Bridge's Baltimore truss form.

Its strong construction is due in part to the fact that it was one of the many post–Flood of 1927 bridges that were built throughout the state, but that is not the main reason. Bridges of the Baltimore truss type were more commonly used on railroads, and this bridge was no exception. It was once a part of the Springfield Electric Railway's trolley line that ran four miles to Charlestown, New Hampshire, providing a short but vital outside transportation link for the town, which was well known for its industrial activity. The railway was established in 1897 and, at its height of activity in 1910, made twelve roundtrips to New Hampshire a day. As with many other railroads, the advent of the automobile doomed the Springfield Electric Railway, though it served until 1947 as the last passenger trolley in Vermont, and its tracks continued to serve freight trains until the early 1980s. Today, the newly refurbished bridge, painted black as many railroad bridges once were, serves highway travel. But is also a part of the Toonerville Trail, a three-mile long recreational trail established by STAG, the Springfield Trails and Greenways citizens group, in the mid–1990s that follows the old Springfield Electric Railway's original line along most of its original length.

---

6. Cavendish — The Depot Street Bridge; DATE BUILT: 1905; TRUSS TYPE: Parker pony (pinned); FEATURE CROSSED: Black River; LENGTH: 98 feet; SPANS: 1; EASE OF ACCESS: Easy

This small one-lane village bridge is important due to one of its design aspects: it is one of only two Parker pinned pony truss bridges remaining in the state, and one of less than ten

The Depot Street Bridge, Cavendish.

pin-connected bridges of all metal truss type bridges remaining in Vermont. Before riveted construction became commonplace by the 1890s, pin connections were used, making bridges of this type more flexible than those that were riveted. Once very common, pin-connected bridges fell out of favor by the 1890s as more rigid bridges were needed to accommodate heavier bridge loads. The Depot Street Bridge is also important as a late example of a pin-connected bridge. All of Vermont's other remaining pin-connected bridges were built in 1900 or prior. The pin connections on this attractive bridge are easy to spot from the pedestrian sidewalk, as are the bridge's other construction details. This bridge is also interesting because of its obscure builder. While many metal truss bridges built in New England after 1900 were built by J.P. Morgan's massive monopoly, the American Bridge Company or an associated contractor, this bridge was built by Henry Norton of Springfield, Massachusetts. This town was an important one in iron production in New England and was the home of the R.F. Hawkins Ironworks concern, a prominent regional bridge builder and the parent company of the only bridge fabricator located in northern New England, the Vermont Construction Company. Though uncertain, Henry Norton was possibly a former employee or officer with Hawkins before starting his own company, and the Depot Street Bridge is his only known bridge in northern New England. This bridge has been recently renovated and, with the removal of the Howard Hill Road truss bridge a few years ago, is the only historic bridge remaining in Cavendish.

7. Hartford — Quechee Gorge Bridge; DATE BUILT: 1911; TRUSS TYPE: Spandrel arch deck; FEATURE CROSSED: Ottauquechee River; LENGTH: 285 feet; SPANS: 3; EASE OF ACCESS: Difficult

The Quechee Gorge Bridge, Hartford.

The Quechee Gorge Bridge stands on its own, so to speak, as one of Vermont's most unusual and historic bridges. However, its spectacular location at one of the state's scenic wonders makes it even more interesting and appealing. The bridge was originally a railroad bridge on the Woodstock Railroad, built to replace a wooden truss that had served since 1875. It was designed by engineer John W. Storrs of Concord, New Hampshire, aided by his son, and partner, Edward Storrs, and fabricated by the American Bridge Company. While Storrs did a great amount of highway bridge design and consultation work in northern New England during the time this bridge was built, he also provided the same for regional railroads as well. Though Storrs was on the payroll of the Boston and Maine Railroad, he also did work for the Woodstock Railroad and the Montpelier and Wells River Railroad in Vermont, to name just a few. This bridge is undoubtedly Storrs' greatest bridge engineering achievement in his long career. A steel arch bridge made sense at this location; it sits some 163 feet above a rocky gorge that is about a mile long and is Vermont's version of the Grand Canyon, carved out by glacial activity over 10,000 years ago. The erection of falsework for a normal truss bridge in this location would have been extremely difficult and costly, while the placement of abutments for a non-arch bridge would have been problematic as well. Instead, the bridge was likely built, though there are no contemporary pictures to document such, with the ribs of the bridge cantilevered outward over the gorge and held in place by cables. Between the steel ribs of the bridge and the road deck above is truss-work that acts as a web to strengthen the bridge. Such a strong bridge was needed because of the heavier locomotives and rolling stock being used by railroads in the 20th century.

The Quechee Gorge Bridge is also significant as the only remaining steel arch bridge entirely within the state of Vermont. It was originally referred to as the Dewey's Mills Bridge, as it was located by the textile mills of the same name. A.G. Dewey moved to the area and established a woolen mill about 1869 that was one of the finest in the country. Well known for its fabric, the mill even made wool used in the uniforms of the Boston Red Sox, the New York Yankees, as well as blankets for the U.S. Army and Navy. Many of these products, in order to get to market, had to cross this notable bridge, first by rail, and then, in 1933, by truck transport when the railroad right of way was taken over for U.S. Route 4 as motorized travel in Vermont was rapidly increasing. To withstand heavy highway traffic, the bridge was strengthened by adding stringers and converting the deck from wood to concrete. Today, the bridge still serves well and carries a heavy amount of traffic on a busy Route 4. As for the woolen mills at this location, they shut down in 1952, with production moved to New Hampshire, and within a few years all the buildings were torn down. The area was soon, by the early 1960s, developed into a state park, and today Quechee State Park is a popular tourist destination. Just as the Quechee Gorge Bridge was vital to mill activities for forty years, it has now had an equally important second life lasting nearly sixty years serving tourists and commuters alike.

---

8. Woodstock — The Elm Street Bridge; DATE BUILT: 1870; TRUSS TYPE: Parker Patent (modified); FEATURE CROSSED: Ottauquechee River; LENGTH: 110 feet; SPANS: 1; EASE OF ACCESS: Moderate

Located near the heart of the historic village of Woodstock, the Elm Street Bridge, with its wrought iron railings, embodies the style and simple elegance ofttimes found in early iron bridges built in more populous areas. Begun in 1869 and completed the following year, this Parker Patent truss bridge is the oldest metal truss bridge remaining in Vermont. It is one of

The Elm Street Bridge in Woodstock.

only three Parker Patent truss bridges, designed by Charles Parker and erected by National Bridge and Iron Works of Boston, ever built in the state. Only five of this type of bridge survive in the United States today, two of them in Vermont, including one also in Northfield. The bridges used wrought iron for all the main members and cast iron for the vertical post connections, which was an innovative feature at the time but one which would prove to be the structures' greatest weakness once heavier loads came along in the early 20th century. As a result, the bridge has been modified with modern bridge components underneath, and now the trusses serve only as a support for its sidewalks. Nevertheless, the bridge has maintained its original appearance and is an exquisite feature on the local scene, serving, as one historian comments, as "a dignified gateway into the village" (McCullough, pg. 114).

---

9. Woodstock — The Mill Street Bridge; DATE BUILT: 1900; TRUSS TYPE: Pennsylvania through; FEATURE CROSSED: Ottauquechee River; LENGTH: 174 feet; SPANS: 1; EASE OF ACCESS: Easy

That the town of Woodstock not only had a need for a great many bridges, but also built a wide variety of them should by now be clear from the preceding entries. The Mill Street

The Mill Street Bridge in Woodstock. The original Pennsylvania truss members form the outermost components of this altered bridge, while the segmented arch within is a modern addition meant to strengthen the structure.

Bridge is, like its neighbors on the Ottauquechee River, unusual for both its truss type, as well as for the strange modifications made in the name of preservation. The bridge is of a Pennsylvania truss design and was built by the Groton Bridge Company of Groton, New York. The development of this truss type came in the field of railroad bridges, but as often happened, spilled over into highway use as well. This bridge is not only one of just three Pennsylvania truss bridges remaining in the state, but also the oldest and the smallest. Though Woodstock today is primarily a tourist town, the Mill Street Bridge, with its strong truss design, harkens back to the days when the town had a large industrial base, manufacturing such things as axes, scythes, carding machines, sashes and blinds, carriages, furniture, leather, and textiles. Bridges in and around such manufacturing concerns needed to be able to handle the heavy loads as raw materials came in and finished products were shipped out, and this bridge did that job very well. Today the bridge serves only a small amount of local traffic and is in reasonably good condition. Sadly, the method chosen for strengthening the bridge was an unfortunate one, and serves to demonstrate that even if a state does have a bridge preservation plan, towns still have control and can make their own choices on bridges located on local roads. The Mill Street Bridge was strengthened with the addition of an arch support system that significantly alters the original appearance of the bridge. While its new appearance is striking due to its bizarre nature, the bridge serves as an example of what not to do when preserving a bridge.

---

10. Woodstock — The Holt Road Bridge; DATE BUILT: 1925; TRUSS TYPE: Warren through, double-intersection; FEATURE CROSSED: Ottauquechee River; LENGTH: 123 feet; SPANS: 1

The Holt Road Bridge in Woodstock, also known as the Bridge Street Bridge.

This bridge, once located just off busy U.S. Route 4 in West Woodstock, was the newest of the four bridges of the multiple-intersection Warren truss type remaining in the state. Unfortunately, it was destroyed by the severe flooding that resulted from Tropical Storm Irene on 28 August 2011. Once a common bridge form, structures with this distinctive light and airy-looking appearance are now a rarity. This bridge, often called the Bridge Road Bridge (though the street is now named Holt Street), was also unusual in that by 1925 few bridges of this type were being built, having been superseded by truss bridges able to bear heavier loads. The Holt Road Bridge served a small amount of local traffic, serving as direct access to a local farm and several residences, and was in good condition prior to its destruction. This bridge will not only be remembered as one of the last examples of an unusual truss type, but also for the beauty it added to the local landscape; the view from Route 4 on a sunny day was stunning, as the bridge almost appeared like a spider-web suspended from the sky. Replacement plans for this bridge, like so many others destroyed in Vermont at the same time, are uncertain as of this writing.

11. Bethel — The Peavine Bridge; DATE BUILT: ca. 1900; TRUSS TYPE: Warren through quadruple intersection; FEATURE CROSSED: Third Branch White River; LENGTH: 102 Feet; SPANS: 1; EASE OF ACCESS: Easy

Like the Paddock Road Bridge in Springfield and the Quechee Gorge Bridge in Hartford, this bridge was formerly a railroad bridge before being converted to highway use in the 20th century when the railroad it served was abandoned. The bridge originally served the White River Valley Railroad (later the White River Railroad), running between Rochester and Bethel for about nineteen miles. This short-line railroad, chartered in 1896 and completed in 1899, was one of many such lines in Vermont during this era. While the railroad suffered many mishaps due to local flooding and poor equipment, ofttimes running twelve hours behind schedule, it would later become more profitable. While the line's equipment was eventually

The Peavine Bridge in Bethel. It once served railroad traffic, but now is a highway bridge.

upgraded, the Peavine Bridge was surely one of its best assets and sound investments early on when it was built, and its survival today is proof of that. The line itself, nicknamed the Peavine Railroad for its meandering course along the White River, had somewhat of a comical reputation among locals. It was said that its train (more like a trolley car early on) ran so slowly that a fellow going to meet his girl could hop off the train, pick a bunch of wild flowers, and easily catch back up with the train. The train was also noted for delivering packages to residences along the tracks, dropping off fishermen at local fishing holes, as well as picking up school children during bad weather...no wonder it constantly ran behind schedule! Among the loads carried on the Peavine were talc and marble quarried in nearby Rochester, as well as lumber and Christmas trees in season.

The line had to rebuild its rickety track numerous times over the years due to flooding, and much of the line, like those of many other railroads in the state, was destroyed during the Great Flood of 1927. The Peavine Bridge seems to have been knocked off its abutments during the flood, but was hauled back into place by horse or oxen. While the line was rebuilt after the Great Flood, it lasted only a few years before going out of service in early 1933. This time it was a combination of the Great Depression, decreased business in local quarries, and an increase in the use of truck transportation that spelled the end of the line. However, one of its finer assets, the Peavine Bridge, lives on. The quadruple-intersection Warren truss structure was common during its time, but is now a rare survivor. Its builder is unknown as the builder's plate is missing from the bridge, but Boston Bridgeworks is one possibility, having built many multiple-intersection Warren truss bridges for railroads in northern New England during this time period. The bridge now serves a small amount of auto traffic on Peavine Boulevard, a short road that follows the old railroad bed for about a mile or so. The area around the bridge is still a popular one with local fishermen.

12. Tunbridge — The Foundry Road Bridge; DATE BUILT: 1889; TRUSS TYPE: Warren pony; FEATURE CROSSED: First Branch White River; LENGTH: 72 feet; SPANS: 1; EASE OF ACCESS: Easy

The Foundry Road Bridge in Tunbridge has been recently restored and is the oldest highway bridge in the state that remains in use (photograph by Daryl Bassett, courtesy of VAOT).

As its name implies, this bridge was originally located next to a blacksmith shop. Smith's Foundry, according to town historian Euclid Farnham, was established in 1855 by Wallace Smith. Through the Civil War, the company made only iron plows, but when Smith's younger brothers, Royal and C. B., purchased the company, the foundry expanded to make stoves sold to area schools and churches, and cultivators. The foundry went out of business by World War I, but the bridge that helped make its expanded business possible remains today as one of the oldest metal truss bridges in Vermont. Quite fittingly, the bridge was fabricated by the Vermont Construction Company of St. Albans, and was the only major bridge in town that was not a covered bridge. It cost the town $876 and replaced an older plank bridge. Though the bridge sees increased traffic today, its setting is still a rural one, with a feed store located adjacent to the site. This bridge, in recognition of its historical status, was renovated in the summer of 2010, though the project was delayed when the bridge trusses were damaged when they were being removed for rehabilitation purposes. The temporary removal of this bridge for several months highlighted the importance of even such small spans as this in the local transportation network, whether in 1889 or the present, as a nine mile or so detour resulted for those trying to reach neighboring businesses and homes.

13. Northfield — The Rabbit Hollow Road Bridge; DATE BUILT: 1908; TRUSS TYPE: Parker pony; FEATURE CROSSED: Dog River and the New England Central Railroad; LENGTH: 88 feet; SPANS: 1; EASE OF ACCESS: Easy

This small bridge is not one of the larger truss bridges remaining in the state, nor is it the oldest. In fact, even in Northfield you can find an older and more historic bridge, the Parker Patent bridge on Vine Street. However, it is small crossings such as the Rabbit Hollow

The Rabbit Hollow Road Bridge, Northfield. Like many metal truss bridges that remain in use in rural areas, it still has a wooden deck.

Road Bridge that have served their local areas long and well with little recognition and fanfare. Once so common, small bridges like this have become increasingly endangered; towns often opt for replacement rather than preserving these seemingly inconsequential structures. This bridge is one of just four Parker pony truss bridges remaining in Vermont, and is the second oldest among them. Located on a back country road, the one-lane Rabbit Hollow Road Bridge, with its steep approach and wooden deck is a hidden treasure for bridge seekers. If you're lucky, while visiting this bridge you might be able to view Amtrak's *Vermonter* passenger train below while it speeds along on daily trips through Northfield on its way north to St. Albans and south to New York and Washington, D.C. The Rabbit Hollow Road Bridge is currently quite rusted and its railing is bent and damaged; work will soon be needed if it is to be saved for future generations.

14. Montpelier — The Granite Street Bridge; Date Built: 1902; Truss Type: Baltimore through; Feature Crossed: Winooski River; Length: 200 feet; Spans: 1; Ease of Access: Easy

This bridge is important for being the oldest and one of just two Baltimore truss highway bridges remaining in Vermont. While Montpelier has four historic truss bridges remaining within its limits today, the Granite Street Bridge is the only one built before the Great Flood of 1927. In fact, this survivor in many ways captures the historic essence of Vermont's capital

The Granite Street Bridge, Montpelier.

city. While Montpelier is the seat of state government, much of its early prosperity was associated with its manufacturing base. Prime among these industrial pursuits was the processing of granite quarried in neighboring Barre. As its name indicates, the Granite Street Bridge is located in an area where the granite sheds were, and still are, located. Here it was that raw granite was hauled over the bridge to be shaped into finished slabs used, among other things, in the construction business, and hauled over yet again when destined for its final locations. Not to be forgotten is the pedestrian aspect of the bridge, as highlighted by the rosettes that decorate the iron railings along its sidewalk; over this bridge granite workers trudged to work every morning, and made the journey back again at night. The Granite Street Bridge has been well cared for over the years and is currently in good condition, though it experiences heavy traffic. However, its location at a difficult intersection, and overall within an expanding urban area makes it potentially vulnerable in the future and a possible candidate for future relocation, as happened recently with the nearby 1928-built Pioneer Street bridge in Montpelier.

---

15. Richmond — The Checkerboard House Bridge; DATE BUILT: 1929; TRUSS TYPE: Pennsylvania through; FEATURE CROSSED: Winooski River; LENGTH: 350 feet; SPANS: 1; EASE OF ACCESS: Difficult

This bridge is the largest Pennsylvania truss span, and the fifth largest pre–1940 metal truss bridge overall in the state of Vermont. Its prominent location on U.S. Route 2 made it an extremely important crossing in the area, and the premier one over the Winooski River in

Vermont Bridges

The Checkerboard House Bridge, Richmond. This view shows its Pennsylvania truss form with the modern interstate bridge above.

Richmond prior to the building of the I-89 interstate, serving as a vital link between Burlington and the Lake Champlain Valley and the state capital at Montpelier. While I-89 has taken over the high-speed traffic as it runs parallel to Route 2 from Montpelier to Colchester, Route 2 remains important as a secondary highway and still experiences heavy traffic. The Checkerboard House Bridge is named for an early 1800s-built house nearby that has prominent Flemish Bond brick work on its end walls. The bridge was built after the Great Flood of 1927 and was one of the largest and most important of the bridges built by the state in the immediate post-flood period in order to rebuild the transportation network and get Vermont moving again.

---

16. New Haven-Weybridge — The Rattlin' Bridge; DATE BUILT: 1908; TRUSS TYPE: Warren through, double-intersection; FEATURE CROSSED: Otter Creek; LENGTH: 148 feet; SPANS: 1; EASE OF ACCESS: Easy

This bridge, as its name implies, was best known in the two towns it served for the noise it made when travelers crossed over, due to its wood planked deck. However, the Rattlin' Bridge is even more historic for its truss form, being one of only four Warren multiple-intersection truss bridges remaining in the state. Of this type of bridge, this one, by far, is the most important in terms of the traffic it serves. While the other multiple-intersection Warren truss bridges in Woodstock and Bethel see but little traffic, and the one in Wilmington

The effort to save the Rattlin' Bridge was both a community and state based initiative that resulted in its full restoration (courtesy of the Vermont AOT).

is bypassed, this bridge is an important link between the towns of Weybridge and New Haven. Prior to its restoration in 2007, the two towns debated for years on whether to demolish the bridge and build a new one, though some citizens did write to the state requesting that the bridge be saved due to its unique nature, and that it not be replaced with a standard modern bridge. Eventually both towns agreed the Rattlin' Bridge was worth saving, and in January 2008 the newly refurbished bridge was formally dedicated. Among those in attendance at the ribbon-cutting ceremony was John Roleau, the great grandson of selectman D.A. Roleau, whose name is on the original bridge maker's plaque as one of the selectmen in 1908 that authorized the bridge's construction. Interestingly, the bridge still made its rattling sound after the bridge was refurbished, but soon had to close when the loose-plank deck was damaged. The wooden deck was subsequently repaired and its planking fixed in place, so the pleasing rattling sound the bridge once made is no more.

---

17. Highgate — The Highgate Falls Bridge; DATE BUILT: 1887; TRUSS TYPE: Lenticular truss; FEATURE CROSSED: Missisquoi River; LENGTH: 295 feet; SPANS: 2; EASE OF ACCESS: Easy

Without a doubt, this bypassed bridge is the most stunning historic metal truss bridge remaining in Vermont today. It is one of only two lenticular truss bridges remaining in the state, and the only one still standing and in its original location (the other formerly stood in Bethel and is now in storage). The Highgate Falls Bridge was originally known as the Keyes Bridge and was built at a cost of $14,000, including $9,400 to the Berlin Iron Bridge Company for the bridge itself, and $4,441 to S.S. Ordway for the masonry work on the abutments, with the state of Vermont paying three fourths of the entire amount. H.W. Varnum of Cambridge, Vermont, was appointed as the bridge commissioner in charge of supervising the job. Interestingly, the site of this bridge is an important one, as it was here that the first covered bridge

The Highgate Falls Bridge, Highgate. This bridge is the largest of the preserved lenticular truss bridges in northern New England and the most beautiful.

in the state was built in 1824. This first bridge at the Highgate Falls site was built by the Keyes brothers, prominent lumber merchants in town, and the name stuck with the new iron bridge for an unknown amount of time. The Highgate Falls are said to be the most powerful in the state. Today a hydroelectric dam is at the site which is owned by the Swanton Village Electric Company, but prior to this Highgate Falls was important for its sawmill and gristmill operations. The bridge formerly stood on a main thoroughfare in Highgate Falls, but it was eventually abandoned and for many years the bridge sat untended and slowly gathered rust.

The Bridge was finally restored in 1999–2000 at a cost of $1 million dollars, the project a joint-venture of the Vermont Agency of Transportation and its Historic Bridge Program, and the Vermont Division of Historic Preservation. It is now a part of a recreational trail developed by the town and is open to pedestrians only. Located off Route 207, the bridge can be difficult to find to those unfamiliar with the area; and it is unfortunate that, at least at this writing, there is no roadside historical marker to guide potential visitors. However, the bridge is well worth any extra time involved in finding it: this two-span, combination truss bridge has a lenticular pony truss for its approach span, followed by a large lenticular through truss spanning the falls. Notable construction details include the pinned connections, the decorative builder's plaque and cresting above, the decorative finials and floral scroll-work on the support columns, and the rosettes on its protective railing. Today, the Highgate Falls Bridge serves not only as a window into the past and the early days of iron bridge building, but it is also a model for historic bridge preservation in the future.

18. Swanton — The Swanton Footbridge; DATE BUILT: 1902, re-erected 2008; TRUSS TYPE: Combination Pennsylvania and Pratt truss; FEATURE CROSSED: Missisquoi River; LENGTH: 447 feet; SPANS: 3; EASE OF ACCESS: Easy

Quite simply, the Swanton Footbridge must stand as the showcase bridge for the state of Vermont's Historic Bridge Program. The two Pennsylvania truss spans of this bridge originally

The Swanton Footbridge, Swanton. The Pennsylvania truss spans, as well as its modern center Warren truss span, are supported by the same piers that once supported the historic covered railroad bridge at this location.

stood in West Milton, Vermont, about twenty miles away from Swanton. However, a series of events would eventually lead to its relocation. The bridge that originally stood at this site in Swanton was a covered railroad bridge built in 1898 on the St. Johnsbury and Lamoille County Railroad, the longest covered railroad bridge in the country. When the bridge was burned in 1987, a historical void was left in the town that would seemingly be hard to fill. Meanwhile, down the road in West Milton, crossing the Lamoille River and measuring 302 feet long, was an aged and rusting Pennsylvania truss bridge built in 1902. The bridge had served well for over 92 years, even surviving the Great Flood of 1927 when many other bridges did not, but now the time for replacement had come.

In 1994 a new bridge was built in West Milton just south of the old bridge, and the state was left wondering what to do about the old bridge. The bridge was eventually dismantled and saved with an eye kept out for a future relocation site: in 2002 the process was started when Scott Newman of the Agency of Transportation called town officials in Swanton and offered them the bridge. Town officials and the Swanton Historical Society jumped at the chance of getting another historic bridge and the reuse campaign began. The Pennsylvania truss bridge was eventually, by 2008, re-erected in Swanton at the same site where the old covered railroad bridge once stood. In fact, the refurbished piers on which the bridge rests originally supported the old railroad bridge. However, the Missisquoi River is a bit wider at this point than the bridge's original location across the Lamoille River, so a solution to span the gap had to be found. This was done with the addition of a 123 foot modern Pratt truss span in the center, which ties together quite nicely the two original spans of the old bridge. The whole relocation project was a massive effort by a great many people, and just goes to show what can be done to save a metal truss bridge when everyone is in agreement! This

Pennsylvania truss bridge is just one of three Pennsylvania truss highway bridges left in the state, and thus was a worthy candidate for preservation. The pedestrian bridge is now a part of a local network of recreational trails and pathways in Swanton, while the location of the Swanton Historical Society Museum just across the street from its western approach offers a great look at the site's historic railroad past.

11

New Hampshire Bridges

1. Nashua — Nashua Manufacturing Company Bridge; DATE BUILT: 1903; TRUSS TYPE: Pratt through; FEATURE CROSSED: Nashua River; LENGTH: N/A; SPANS: 1; EASE OF ACCESS: Easy

The overhead bracing system of the Nashua Manufacturing Company Bridge combines both strength and beauty in industrial design.

This bridge is one of the hidden treasures of this industrial city's history. Though it is tucked away behind the former Nashua Manufacturing Company (NMC) mill buildings, now called Clocktower Place, it has been renovated and remains a unique feature among the millworks it once served. The NMC was one of the first industrial concerns in Nashua, was a key factor in the city's early growth and prosperity, and was also the first large textile mill in New

Hampshire. Founded in 1823, the NMC remained in operation until closing in 1948. The cotton textile operations of the mill were noted both regionally and nationally during the company's lifetime, and it was a key supplier to the U.S. military during World War II. The bridge was important within the complex in transferring raw materials via shuttle-cars and carts and is an important survivor, as few of the metal truss bridges utilized in factory and mill buildings in northern New England have survived to this day. Upon the shuttering of the NMC mills by its owner, Textron, in 1948, the complex and the bridge itself languished for many years. As with many former mill buildings in New England, those of the old NMC have been renovated and turned into upscale apartments since 2000. The bridge itself has been renovated and now serves pedestrian traffic, though only one end of the bridge is accessible to the public.

---

2. Milford — The Swinging Bridge; DATE BUILT: 1889; TRUSS TYPE: Suspension; FEATURE CROSSED: Souhegan River; LENGTH: N/A; SPANS: 1; EASE OF ACCESS: Easy

The Swinging Bridge, Milford. This view shows the bridge's tower and the cables it supports.

This picturesque bridge, and rare survivor, is known locally as the Swinging Bridge for the motion its makes as one passes over. Simple pedestrian suspension bridges of this type were once common in northern New England and were utilized in a variety of settings, including factory and mill settings, as well as centerpieces in small towns and villages. The builder of this bridge was the prolific Berlin Iron Bridge Company of East Berlin, Connecticut. Salesmen for this company were adept at selling their bridges to local town officials: long after these local selectmen have been forgotten, their names remain on bridge maker's plaques as the officials in charge when the decision to build the bridge was made, perhaps one of their finest, and surely longest-lasting, achievements! The Swinging Bridge offers its pedestrians a first hand look at the simple mechanics of a suspension bridge: note the anchored cable-stays in the ground as you approach the bridge, the overall key to keeping the bridge in place. As you get closer, your eyes can't help but be drawn upward to the tops of the twin iron pylon supports at each end of the bridge. While you may be pleasantly distracted by the elegant builder's plaque cresting above the tower supports and decorative finials atop each pylon, be sure to pay attention to where the wire cables that suspend the bridge pass through the tower support saddles before dipping gradually downward to join with the bridge deck that it supports. This miniature Golden Gate Bridge, which appears to be in good condition, once served factory workers and those out for a casual stroll as well. Today, it is an important historical link to the pre-automobile past!

3. Milford — The Jones Crossing Bridge; DATE BUILT: 1910; TRUSS TYPE: Pratt through; FEATURE CROSSED: Souhegan River; LENGTH: 150 feet; SPANS: 1; EASE OF ACCESS: Easy

The Jones Crossing Bridge, Milford.

While Milford's Swinging Bridge may hearken back to a day when people walked or rode a horse to get from place to place, the Jones Crossing Bridge just across town is symbolic of the day when the car was becoming king. This riveted structure, located off busy state Route 101A, has been bypassed since 1992 and now sits rusting and forlorn. However, once upon a time it was an important link in the rural road system in the western part of town. The bridge is also important for its association with engineer John W. Storrs who designed the crossing, as well as its builder, the Canton Bridge Company of Canton, Ohio. This company was a successor to the Wrought Iron Bridge Company and was a prolific builder in the Midwest and elsewhere on the East Coast, but this is their only known survivor among northern New England bridges. It is interesting that the Milford selectmen chose this company, no doubt getting an attractive price, at a time when the American Bridge Company often steamrolled over its competition. The future of this slowly deteriorating bridge is uncertain at this point. While ten pre–1940 riveted single-span highway Pratt through truss bridges remain in the state, this one is the oldest.

---

4. Exeter — Park Street Bridge; DATE BUILT: 1892; TRUSS TYPE: Warren pony; FEATURE CROSSED: Boston and Maine Railroad; LENGTH: 58 feet; SPANS: 1; EASE OF ACCESS: Easy

The Park Street Bridge, Exeter.

This seemingly simple and unprepossessing structure is actually one of the most historic metal truss bridges in New Hampshire. Its riveted construction, later common, was rare and expensive in 1892 when pin-connected structures were the norm, while the lattice diagonal members were also unusual. In addition, the Warren pony truss was once one of the most commonly built bridge forms throughout New England, but it is now endangered. Their simplicity has often led to their quick replacement with little or no preservation considerations when a new bridge was needed. In a 1985 survey, New Hampshire had 26 pre–1940 highway bridges of this type remaining, but today there are only nine, two of which are closed. Because of this great loss, about 65 percent thus far, Warren pony truss bridges in the state have been given increased attention with an eye toward stemming these losses. The Park Street Bridge in Exeter is also notable for its important regional builder, the R.F. Hawkins Ironworks of Springfield, Massachusetts. This bridge fabricating company was well-patronized by the Boston and Maine Railroad for a number of their small crossings, but few remain in existence in northern New England today. The future of this historic bridge is perhaps uncertain, but its location in a neighborhood setting with relatively light and slower moving traffic over a still-active rail line may help its cause.

---

5. Hancock-Greenfield — Cavender Road Bridge; DATE BUILT: 1905; TRUSS TYPE: Pratt pony (pin-connected); FEATURE CROSSED: Contoocook River; LENGTH: 75 feet; SPANS: 1; EASE OF ACCESS: Easy

The Cavender Road Bridge, Hancock-Greenfield.

This now abandoned bridge was built by the Groton Bridge Company of New York. The town of Hancock paid the bridge company $433 for their share of the project and, though uncertain, it seems likely the town of Greenfield paid a similar amount, making the bridge's total cost somewhere in the neighborhood of $850. The bridge for many years was known as the Dennis Bridge, for a landowner in the area, or, as was quite common, simply "the Iron Bridge." It is interesting to note that both towns plowed up to the bridge during winter snow storms, but neither town plowed the bridge itself for many years! Why? When the bridge was built, many New England towns commonly used large, horse-drawn wooden snow rollers to pack down the snow to help facilitate sled travel, especially through covered bridges. However, within a few short years after the bridge was built, the town converted to motorized tractor plows which were both too heavy and too wide for the relatively new bridge! Despite this oversight, any doubt the town of Hancock might have had in expending money to build newer truss bridges was probably dispelled in 1912, when local farmer William Welch met with an accident on a town bridge when it collapsed while he was carting a load of apples. Welch sued and eventually settled for the tidy sum of $2,500 — some things never change!

Though the Cavender Road Bridge is small, it is an important survivor: it is now the sole remaining pin-connected Pratt pony truss bridge in New Hampshire, and one of just eight pin-connected bridges remaining in the state. Take a good look if you visit this bridge in person. The pin connections may be easily viewed, and if you look close enough, you can see that the bridge's steel beams are stamped with the company that forged them, the Jones and Laughlin Steel Company of Pittsburgh. Though the bridge is abandoned today and its future uncertain, it is in relatively good condition and lies in a forested, almost park-like setting that might possibly lend itself to be turned into a real park that could be jointly administered by both Hancock and Greenfield.

---

6. Antrim-Bennington — The Thompson Crossing Bridge; DATE BUILT: 1893; TRUSS TYPE: Pratt through (pin-connected); FEATURE CROSSED: Contoocook River; LENGTH: N/A; SPANS: 1; EASE OF ACCESS: Difficult

The Thompson Crossing Bridge, though abandoned and slowly falling to pieces, retains a geometric elegance in a heavily wooded setting.

The Thompson Crossing Bridge is important as one of the oldest pin-connected Pratt truss bridges in the state, built in the same year as the only other bridge of its type remaining, the Stratford, New Hampshire-Maidstone, Vermont, bridge over the Connecticut River. However, while the latter has been lovingly restored, this bridge has been abandoned. The Thompson Crossing Bridge has been closed for about two decades and lies on an abandoned and badly rutted road between Antrim and Bennington in a heavily wooded area. The bridge now has no deck and it is only a matter of time before it falls into the river below. In 1997 the town requested an estimate for its replacement, but nothing further has been done and the chances for saving the structure seem slim.

---

7. Chichester — Thunder Bridge; DATE BUILT: 1887; TRUSS TYPE: Lenticular through truss; FEATURE CROSSED: Suncook River; LENGTH: 96 feet; SPANS: 1; EASE OF ACCESS: Easy

The Thunder Bridge, Chichester.

When this bridge was built, it signaled a new era for the town of Chichester and great celebrations took place at the opening of the impressive structure. Over the years, the bridge, built for $1,950 and sometimes also called the Pineground Bridge, served not only pedestrians, and later, auto traffic; but it was also a place where couples could take a stroll and was a great location from which to fish and swim. It truly was a town landmark. However, the bridge was built for a different age, and it eventually fell into disrepair and was no longer suitable for auto travel. Thunder Bridge was bypassed in 1981 when a new bridge was built, and languished for over twenty years. However, the town recognized the importance of the bridge, the only lenticular through truss structure remaining in the state, and with the help of the New Hampshire Division of Historical Resources, it was restored in 2005. Today, the bypassed bridge has been turned into a small park, with signing and community information that highlights its history. In many ways, the preservation of Thunder Bridge serves as a shining example of bridge preservation at its best. This bridge was built by the Berlin Iron Bridge Company of Connecticut, the most prolific builder of this type of bridge. Their salesmen blanketed New England and built hundreds of this kind of bridge, but what was once common is now a rarity. The pin connections on Thunder Bridge are readily visible, while the finial-topped end posts and decorative builder's plaque and decorative cresting are hallmarks of a Berlin Iron Bridge Company structure.

---

8. Dover-Newington — The General Sullivan Bridge; DATE BUILT: 1934; TRUSS TYPE: Continuous Steel truss; FEATURE CROSSED: Piscataqua River, Little Bay; LENGTH: 1,585 feet; SPANS: 9; EASE OF ACCESS: Difficult (temporarily closed winter 2010)

The General Sullivan Bridge, Dover Point–Newington.

The General Sullivan Bridge is one of the most historic metal truss bridges in the state, located in an equally historic area. Its location in the Dover Point-Newington area was the site of one of the first settlements in the New Hampshire colony, being continuously settled since 1623. Several historic bridges were built in this area over the years, including the 1797-built Great Arch Bridge, a 2,362 foot long wooden bridge, and later an 1873-built 1,648 foot long combination bridge that consisted of several wooden pile and beam spans, a covered Howe truss span, and a swing bridge span for local boat traffic between Little and Great Bay and the Piscataqua River. This combination bridge served until 1934 when it was replaced by the General Sullivan Bridge.

Though the bridge, the only continuous truss bridge in the state today, is often regarded as an eyesore and a rusting hulk, the General Sullivan was one of the most advanced structures of its day. It employs a continuous truss design that was gaining in popularity in the 1930s for long-span bridges, and was engineered by the firm of Fay, Spofford, and Thorndike of Boston, the bridge's principle designer being Charles M. Spofford, a noted expert on continuous truss bridges. The General Sullivan is also notable for being a WPA (Works Progress Administration) project for the New Hampshire Toll Bridge Commission, and was one of the most significant Depression Era projects in the state financed as a result of President Roosevelt's New Deal administration. A plaque on the bridge even documents its status as Public

Works Project docket number 752. The bridge was fabricated by the Lackawanna Steel Corporation, a subsidiary of the Bethlehem Steel Company, and consists of nine spans measuring 102, 175, 163, 200, 275, 200, 163, 163, and 125 feet individually for a total length of 1,585 feet. All of the spans but the center are of the deck truss type, while the center through truss arch span is the largest, measuring 275 feet in length. This portion of the bridge was significant as it allowed boat traffic to freely pass through underneath, a marked improvement over the old swing bridge span. The General Sullivan was built as a toll bridge, with motorists paying a ten cents toll in the beginning, and remained a toll bridge until it was bypassed in 1984, being replaced by the Little Bay Bridge, which was built right next to the General Sullivan beginning in 1966. The first span of the new bridge (now the southbound span) was opened in that year, while increased traffic in the area resulted in a second span (the northbound side) opening in 1984.

With this bridge complete, the General Sullivan Bridge was closed to motor vehicle traffic, but remained open to pedestrian and bike traffic. Since 1984 the General Sullivan Bridge has remained popular with walkers, joggers, fishermen, and even local commuters. Indeed, the bridge is vital to local residents that work on the other side of the river, as the Little Bay Bridge is too narrow and too congested to accommodate any pedestrian traffic. However, the General Sullivan has had only scant maintenance over the years and it is now badly rusted and its concrete deck continues to deteriorate. For years the bridge's fate was debated. Many thought the bridge should be demolished, including the U.S. Coast Guard, which regards the bridge as a navigation hazard. Yet others recognize the structure's historic status and wanted to see the bridge refurbished. Since the four-lane Little Bay Bridge cannot now handle the volume of traffic the highway carries, various options have been debated which included either the demolition of the General Sullivan so that the Little Bay Bridge can expand to eight lanes, or the refurbishing of the General Sullivan so that it can handle several lanes of traffic, with the Little Bay Bridge expanding to six lanes.

While the debate has been ongoing for over a decade, it now appears that a new bridge will be built between the Little Bay and General Sullivan bridges, and that the General Sullivan will be rehabilitated and kept in its current role as a pedestrian bridge. The total cost of the project, which began in September 2010, will be about $240 million, with the refurbishing of the General Sullivan Bridge expected to cost about $26 million. However, the bridge's future is not yet totally assured, as the state of New Hampshire has commented to the effect that if hidden structural defects are discovered which might substantially increase the cost of fixing the bridge, the option of demolishing it might have to be revisited. While only time will tell what the fate of the General Sullivan Bridge might be, there is cause for concern as similar landmark in northern New England, the Lake Champlain Bridge (also a continuous truss bridge designed by Fay, Spofford, and Thorndike), was demolished in 2009 due to defects that were too costly to fix. Hopefully this will not prove to be the case with the General Sullivan Bridge. Already a prominent feature on the local landscape, a coat of paint and a thorough going-over might just restore the bridge to its former glory.

---

9. Hooksett — Lilac Bridge; DATE BUILT: 1909–1936; TRUSS TYPE: Pratt through; FEATURE CROSSED: Merrimack River; LENGTH: 491 feet; SPANS: 3; EASE OF ACCESS: Difficult/Restricted

This pleasantly named bridge occupies an important and historic site in the village of Hooksett, just north of Manchester. Lilac Bridge occupies the site of the old village railroad

The Lilac Bridge (right), Hooksett.

bridge (the pedestrian covered bridge was located just upstream), while just downstream is a steel truss railroad bridge that is still in use on one of the oldest railroad lines in the state. At the village end of Lilac Bridge sits Robie's Country Store. Originally founded by George Robie, the store was a part of daily life in Hooksett for over 100 years, being operated by the Robie family and its descendants from 1887 to 1997. Still in operation today as one of the state's best preserved country stores, Robie's has also been prominent for years as a meeting place for national politicians. Because of its first in the nation presidential primary election every four years, national candidates have made the trip to New Hampshire to test the waters, and one of the most visited spots for retail politicking over the years has been this local store.

The Lilac Bridge is the only remaining three-span Pratt truss bridge in the state, and has seen its share of damage over the years due to the raging flood waters of the mighty Merrimack River. During the terrible flooding of 1936, Hooksett was hard hit, and the southern most span of Lilac Bridge was destroyed, rebuilt that same year by the American Bridge Company. The Lilac Bridge is also significant because it was designed by John Storrs, New Hampshire's premiere bridge architect of the early 20th century. However, as with many early truss bridges, the Lilac Bridge was too narrow for increased local traffic and it was bypassed in 1976. Ever since that time it has been neglected and has slowly deteriorated to the point where the bridge is now totally closed and fenced off at either end. While options to rehabilitate the bridge as a pedestrian crossing have been discussed over the years, nothing has yet been done. While many agree the bridge is worth saving, its great size, and the huge amount of funds needed to refurbish it make the future of the Lilac Bridge uncertain.

10. Dover—Broad Street Bridge; DATE BUILT-1901; TRUSS TYPE: Baltimore; FEATURE CROSSED: Broad Street; LENGTH: N/A; NUMBER OF SPANS: 1; Ease of Access-Difficult

The Broad Street Bridge, Dover.

This bridge, formerly on the Boston and Maine Railroad, is one of the oldest railroad bridges in the state. Its builder is unknown, though it appears to have been built by the Phoenix Bridge Company of Phoenixville, Pennsylvania. The bridge has long been a part of the cityscape of Dover, and it's a bridge that many in town love to hate. As a former resident of this town for nearly a decade, I must confess that I have an affinity for this bridge. Many, however, do not, mostly truck drivers and other large vehicle operators, since the clearance for going under this bridge is less than ten feet. In fact, I nearly made that mistake myself while driving through town in a moving van, remembering just in time so that I had to come to a stop and back up in the middle of the street, disrupting traffic. Oft-times, however, drivers fail to pay attention and more than once during my time in Dover I saw trucks either stuck underneath or stopped because their tops have been ripped off or damaged as they tried to make it through. This bridge also poses problems for local fire vehicles. The Broad Street Firehouse is located within sight of this bridge, but when making calls to the east of town often has to detour around the bridge. Despite all this, the bridge is historic and its elevated setting offers a pleasing view of its skewed trusses, especially at sunset. The Broad Street Bridge is also an active one and if you're a rail-fan, this bridge is just for you; Amtrak's *Downeaster* uses the tracks five times a day and makes a stop just a short distance away at the station, while diesel freight trains also rumble through on a daily basis.

11. Claremont — Monadnock Mills Footbridge; Date Built: ca. 1870; Truss Type: Bowstring Arch; Feature Crossed: Sugar River; Length: N/A; Spans: 1; Ease of Access: Moderate

The Monadnock Mills Bridge, Claremont.

This factory bridge, located at Monadnock Mills #6, is extremely important not only as the oldest metal truss bridge in New Hampshire, but the only standing Moseley Iron Bridge Company bridge in all of northern New England. In fact, only five Moseley bridges are known to be in existence, one in storage in Bennington, Vermont, one in North Andover, Massachusetts (formerly a factory bridge), one in Kimberton, Pennsylvania, and one on display at the Henry Ford Museum in Dearborn, Michigan. This bridge company was established by Thomas Moseley in Cincinnati, Ohio, later moved to Boston and was a prolific builder of his patented wrought iron arch bridge from 1858 to 1879.

The bridge in Claremont was formerly a footbridge for workers between the mill and the gasworks across the Sugar River, but now it has no deck and serves as a simple pipe chase. While the mills are now being renovated after having been empty for many years, it is hoped that this bridge will soon be renovated and get the attention it deserves. Originally chartered as the Sugar River Manufacturing Company, with waterpower rights for the location granted in 1831, it was not until 1844, under its new name, that the Monadnock Mills were established and in operation. This cotton textile mill became one of the first and was once the largest mill operations in the upper Connecticut Valley, employing at its peak about 500 workers, producing over 2 million yards of cloth a year and over 90,000 quilts. The company went out of business in 1963, and many of its local buildings were sold to private interests. Today, the city of Claremont owns the Monadnock Mills buildings and has worked hard to revitalize the complex by finding new uses for its building, thus saving this historic area while at the same time bringing in some new jobs and furthering the town's economic growth. This iron footbridge is not currently open to the public, but is easily viewable from several public areas around the mill complex and near the site where the gasworks once stood. The triangle shaped riveted tubular arches of the bridge are a pioneering design patented by Moseley and are clearly visible and can be studied close-up with binoculars or a telephoto lens, as can be the small builder's plaque mounted at the top center of the bridge.

12. Claremont — Sullivan Machinery Company Bridge; DATE BUILT: 1905; TRUSS TYPE: Baltimore through; FEATURE CROSSED: Sugar River; LENGTH: 160 feet; SPANS: 1; EASE OF ACCESS: Moderate

The Sullivan Machinery Company Bridge, Claremont.

Of the three states in northern New England, New Hampshire has the largest number of the old factory and mill-related metal truss bridges remaining. This bridge is one of them, and is certainly one of the most impressive in terms of size and the important factory it served. Employing a Baltimore truss, the SMC Bridge was used by workers and shuttle cars to transport materials and finished machinery to and from the factory complex on both sides of the river. Founded by J.P. Upham and Albert Ball, the SMC was formed in 1868, manufacturing a patented diamond channeling machine used in the mining industry. With the profits made, the SMC complex expanded greatly in the 1880s, and by 1890 the company was a pioneer in the mining industry, inventing the first electric-driven diamond core drill and several different types of coal cutters. These products were used the world over, and the small New Hampshire company soon merged with the Diamond Prospecting Company and the main office was located in Chicago. However, the New Hampshire operations remained as the primary production facilities and the company continued to grow, reaching its business peak in 1903.

This bridge was built by the American Bridge Company to connect the old factory on the south side of the Sugar River to the newly built complex on the north side. This purchase of waterfront property where a foundry was soon established was at the time the largest real-estate deal ever in Claremont. Indeed, SMC was Claremont's largest employer for many years. The mill complex, largely abandoned and having sustained some fire damage, is now owned by the city and is part of their massive Claremont Rehabilitation Project. Just what role the historic bridge will play remains yet to be determined, but it and the factory it served are proud reminders of a time when mining equipment from this town was shipped all over the world.

---

13. Concord — Sewall's Falls Bridge; DATE BUILT: 1915; TRUSS TYPE: Pratt through; FEATURE CROSSED: Merrimack River; LENGTH: 335 feet; SPANS: 2; EASE OF ACCESS: Easy

The Sewall's Falls Bridge, Concord.

This bridge is one of those spans that serves as a poster-child, so to speak, for the plight of metal truss bridges everywhere, both good and bad. Its historical association with John W. Storrs, the noted civil engineer and later mayor of Concord, is important as it is only bridge of his remaining in Concord. Proposals have even been made to rename the bridge in his honor. While Storrs, along with his son Edward, designed the bridge, it was built by the Berlin Construction Company. On the other hand, however, there are those locals who despise the bridge, perhaps with equally good reasons; the Sewall's Falls Bridge is rust-streaked and far from pleasing to the eye. Even worse, the bridge is somewhat dangerous. Built during an age when autos were much smaller and fewer in numbers, the bridge is now essentially a one-lane bridge. Those who use the bridge daily have to take their turn and risk a crossing, even when the intentions of traffic at the far end of the bridge are not always clear. It is a difficult situation, and one referred to locally as playing a game of Sewall's Falls chicken!

Interestingly, discussions have taken place over the years to either demolish the bridge or bypass it and build a new one, but nothing has yet been decided. The local press, the *Concord Monitor*, has commented against the bridge for a number of years now, but with an acknowledged bias: its headquarters are located on the eastern side of the bridge and employees (i.e., reporters and editors) have to use the Sewall's Falls Bridge daily. As dangerous as this bridge is, it once had an even more difficult and steep western approach. This situation was remedied in 1937 when a new approach was built consisting of seven I-beam and concrete spans. The steel beams were manufactured by Bethlehem Steel Company, while the approach ramps were built by the Simpson Brothers Corporation, all part of a WPA (Works Progress Administration) project. The Sewall's Falls Bridge is the oldest and one of just two twin-span Pratt through bridges remaining in the state. The other one was also designed by Storrs and once stood in Concord until its removal and re-erection in Henniker on Western Avenue in 1933.

14. Boscawen-Canterbury — The Depot Street Bridge; DATE BUILT: 1907; TRUSS TYPE: Parker through; FEATURE CROSSED: Merrimack River; LENGTH: N/A; SPANS: 2; EASE OF ACCESS: Difficult/Restricted

A portal view of the abandoned Depot Street Bridge.

This now abandoned bridge was built by the United Construction Company of New York at a cost of $6,550 and was likely fabricated by the American Bridge Company, with whom United typically partnered. The bridge was designed by John Storrs, for which he was paid $163.75. Storrs was also employed by the town in 1907 to inspect all their bridges, doing so for $15! The Depot Bridge had some big shoes to fill as it replaced the legendary Rainbow Bridge, a McCallum truss covered bridge that served on the old road to Canterbury for 50 years before being smashed to pieces during the flood of 1907. That the new steel bridge did its job well is evident by the fact that it survived the devastating floods of the 1930s. The bridge was once important, being located off U.S. Route 3, the main highway north out of Concord before the interstate was built, on a road that crossed the Merrimack River to the neighboring town of Canterbury.

Once interstate I-93 was built in the 1960s north out of Concord and through Canterbury, with an exit just east of the bridge, the road between Boscawen and Canterbury was much reduced in travel. The bridge subsequently fell into disrepair and was closed in 1965, a victim of changing traffic patterns. The Depot Street Bridge is now in poor condition and fenced off for safety reasons, and it is unlikely it will ever be rehabilitated in its original location, even though the spot would seem to lend itself to preservation as a pedestrian bridge as part of a recreational trail. In April 2010 the possibility of dismantling the bridge with an eye toward re-erecting it elsewhere was discussed, but no decisions has yet been made. The bridge is one of just three remaining twin-span Parker truss bridges remaining in New Hampshire, and is by far the oldest.

---

15. Ashland-Bridgewater — Pemigewasset River Railroad Bridge; DATE BUILT: 1903; TRUSS TYPE: Warren; FEATURE CROSSED: Pemigewasset River; LENGTH: 410 feet ;SPANS: 3; EASE OF ACCESS: Difficult

An early view of the Pemigewasset River railroad bridge, showing a Boston and Maine steam locomotive making a crossing (courtesy of Ashland Historical Society).

This railroad bridge was built to replace a wooden covered railroad bridge on the Boston and Maine Railroad that burned in January 1902. The area in which this bridge stands has seen a lot of bridge building activity over the years, with highway and railroad bridges crossing in close proximity to one another for well over a hundred years now. Today, the railroad bridge is clearly visible from the 1931-built Pratt deck truss bridge that carries U.S. Route 3 over the Pemigewasset River just upstream. Like so many bridges built in the early 1900s in New Hampshire, the Pemigewasset River Railroad Bridge was designed by engineer John Storrs. Newspaper accounts state that 700,000 pounds of steel, with a tensile strength of 60,000 pounds per square inch, were used in the construction of the bridge and that it is held together by 15,000 rivets that were hand driven. These same accounts also mistakenly identify the bridge as a Pratt truss structure. The steel for the bridge was shipped to the site in early December 1902 and construction began straight away. The first span was completed by mid–January 1903, and the remaining two spans in February, with final completion by March 1903. Interestingly, the granite piers from the former covered bridge were utilized for the new steel bridge, and the only wood used in the entire structure was for the railroad ties themselves. The bridge was an active part of the Boston and Maine Railroad, but was reduced to freight service when passenger service to Ashland ended in 1959.

The bridge and the old Boston and Maine line from Concord to Lincoln was purchased by the state in 1975. Freight service used to run on this line even after passenger service ended, serving the Ashland Paper Company and the Rochester Shoe Tree Company, located just before the southern entrance of the bridge, and other local businesses into the 1970s. Locals recall that when the shoe tree company was active, the smell of cedar was strong in the bridge area because of the cedar wood stored by the company. Trainmen will also tell you that remains of the accident that ruined the old covered railroad bridge can sometimes be seen from the current railroad bridge: the covered bridge burned when a double-header freight train and a gravel train collided head on in the center, sending three locomotives and a number of freight cars into the Pemigewaset River below and the remains of an upside down flat car can still be seen in the water at certain times. The line on which the bridge serves is today leased from the state by the Lincoln-based Hobo Railroad and sees occasional seasonal passenger service.

---

16. Campton — Livermore Falls Bridge; DATE BUILT: 1885; TRUSS TYPE: Lenticular deck; FEATURE CROSSED: Pemigewasset River; LENGTH: 263 feet; SPANS: 2; EASE OF ACCESS: Very Difficult

A view of the remaining span of the abandoned Livermore Falls Bridge, Campton (courtesy of Campton Historical Society).

The iron skeleton of its largest span is all that remains of the Livermore Falls Bridge. Though the iron stringers for the bridge's deck are still in place, the mill it once served is now in ruins and the area has largely reverted to wilderness.

This bridge is not only one of the oldest metal truss bridges in New Hampshire it is also one of the truly hidden treasures among northern New England bridges. The bridge is known locally as the Pumpkin Seed Bridge due to the shape of its lenticular truss, and is part of a ruined mill complex dating back to the 1880s. The Livermore Falls area was first settled in the 1770s by Moses Little, who established a sawmill and a gristmill and was a prominent official in town. The falls, however, gained the name by which it is known today from a later settler, Judge Arthur Livermore, who acquired the property and water rights to the area in 1827. Additionally, the Livermore Falls area was also the site of the first fish hatchery established in the state, going into operation in 1788.

As to the current bridge, it replaced a bridge built in 1869 that appears to have been a wooden truss structure and whose expense was shared by the towns of Holderness, Plymouth, and Campton. Several pulp mills were established at Livermore Falls, the first in 1888, and a much larger one further upstream the following year. The ruins of this mill, which remained in operation until the 1950s, are close by the western approach to the Livermore Falls Bridge, in a wooded area off U.S. Route 3. Within a few years of the pulp mill closing, the Livermore Falls Bridge was closed in 1959. Ever since then, the bridge has deteriorated due to the effects of time and mother nature. A flood in 1973 destroyed the dam and powerhouse of the mill complex, while the smaller eastern span of the bridge collapsed many years ago. The deck of the remaining span has long since rotted away and now just the iron skeleton of the bridge remains. Even in its current ruined condition, the bridge is an awe-inspiring site, perched 103 feet above the Pemigewasset River. Attempts at revitalizing the area came in the late 1980s, when the building of a hydroelectric dam was proposed, but this effort fell through when the necessary permits were denied. The area on the eastern side of the bridge was designated the

Livermore Falls State Forest by the late 1980s. The Livermore Falls Bridge is important due to its historic setting and lenticular deck truss design, having been built by the Berlin Iron Bridge Company of Connecticut, but it seems unlikely that the bridge will ever be restored to its original state. The bridge is very difficult to access due to the steep and rocky slopes on each side of the river, but in spite of this the bridge and the ruined factory have been for many years a favorite destination of local hikers and explorers.

---

17. Wentworth — Wentworth Village Bridge; DATE BUILT: 1909; TRUSS TYPE: Warren through; FEATURE CROSSED: Baker River; LENGTH: 97 feet; SPANS: 1; EASE OF ACCESS: Easy

The Wentworth Village Bridge, Wentworth.

One of the major aesthetic differences, it might be argued, between covered bridges and metal truss bridges is that the former seems to blend in with the local landscape, while the later sometimes stand in stark contrast to nature. Not so with this bridge; tucked away on the village road within view of the town meetinghouse spanning the Baker River above the Cascading Falls in a wooded setting, this bridge seems to fit right in. The Wentworth Village Bridge was one of a number of bridges built in New Hampshire after the 1907 flood and was designed by John Storrs. While the bridge was bypassed in 1994 and is now open only to foot traffic, its setting was not always so quiet. The village area was once a beehive of activity, home to Cole's Carriage Shop and the Knight and Crosby Bobbin Factory. Outstanding features of the Wentworth Village Bridge, beyond its setting, are its decorative iron railings and builder's plaque. It is, however, in need of some attention; a coat of paint would restore it nicely, while its wooden deck will soon need replacing. The bridge is one of eight single span Warren through truss bridges left in the state, two of which are in Wentworth, and the oldest. All but one of the other bridges of this type are post-flood bridges built in 1927 to 1928 or 1937, the exception being the other historic bridge in Wentworth, the Sanders Hill Road Bridge, built by the American Bridge Company in 1930.

18. Tilton — Tilton Island Park Bridge; DATE BUILT: 1881; TRUSS TYPE: Truesdell; FEATURE CROSSED: Winnipesaukee River; LENGTH: 83 feet; SPANS: 2; EASE OF ACCESS: Easy

The Tilton Island Park Bridge, Tilton.

The site of this historic footbridge is unique and one that is tied directly to Tilton's greatest benefactor. Charles Tilton was the grandson of town founder, Nathaniel Tilton, and was perhaps the wealthiest man in town. Not only was he a mill owner, but he was also a world traveler and a man of grandiose ideas. He donated many unique statues to decorate the town, and even commissioned the building of the Memorial Arch in 1882–1883, modeled after the Arch of Titus in Rome and today the town's most recognized landmark. That Charles Tilton was inspired by the classic arch in many of his architectural pursuits is further evidenced by one of his other long-lasting contributions to his hometown, the Tilton Island Park Bridge. Tilton Island itself is largely manmade; fill from the construction of the railroad and the rebuilding of a mill canal was added to a small island to make it a more substantial location. The island served a variety of uses early on, but came to prominence after it was bought by Charles Tilton in 1865. He converted it into an elegant park, filling it with statuary and even a fancy summerhouse. To get to the park, Tilton purchased a bridge that was to be equally elegant. Tilton was a busy man in 1881: not only did he visit Rome that year and, once he arrived home, began the planning of the Memorial Arch, he also commissioned the Tilton Island Park Bridge.

The bridge's builder was A.D. Briggs of Springfield, Massachusetts, a former partner of the R.F. Hawkins Ironworks concern who started his own bridge company in 1864. How Briggs came to the attention of Charles Tilton is unknown, but the form utilized by this rare cast iron bridge, the Truesdell truss, was a most unusual choice. This truss, consisting of an iron lattice system, was patented by Lucius Truesdell of Warren, Massachusetts, in 1858 and was used in the Midwest. However good the truss design may have appeared on paper (in the days before bridge designs could be stress tested), it was a deadly failure for long span bridges. One Truesdell truss bridge built in 1866 over the Fox River in Elgin, Illinois collapsed in 1869 when a large amount of cattle, nearly 100 head, were crossing. While no one was killed in this collapse, just a few years later, in 1873, a four-year-old Truesdell truss bridge over the Rock River at Dixon, Illinois, collapsed, killing 46 people and injuring many more. With this accident, the Truesdell truss was largely a discredited design and faded from history—well, almost. Albert Briggs of Springfield, Massachusetts was an early agent for Truesdell truss bridges, and apparently kept building them even after these failures, albeit on a much smaller scale.

The light nature of the Tilton Island Park Bridge is evident upon first view, its six panel trusses are almost lacey in appearance from a distance, and close-up resembles more the wrought iron railings of later metal bridges! A further indicator of the Truesdell truss and its inherently weak design is the presence of a center pier for a bridge well under 100 feet long. While I've not been able to determine if the center pier was an original feature of the bridge, though it is depicted in postcards a century old, its use in any case is indicative of the additional support required by the bridge. The Tilton Island Park Bridge significance lies in the fact that as the only known Truesdell truss bridge in New England, and possibly the only one left in the country; it is a truly rare bridge type. Today, this arched lattice bridge is well cared for by the town of Tilton and is one of its many unique landmarks.

---

19. Franconia — The Delage Bridge; DATE BUILT: 1889; TRUSS TYPE: Lenticular pony; FEATURE CROSSED: Creek outlet to Gale River; LENGTH: Approx. 60 feet; SPANS: 1; EASE OF ACCESS: Easy

Founded along the banks of the Gale River and its many tributary brooks, the town of Franconia has always had a vital need for small bridges to keep the town connected. While its early bridges were, of course, made of timber, the town turned to iron bridges by the 1880s. One might say the town had a true appreciation for the potential of this new bridge material. Franconia itself, with its local deposits of iron ore, was important for producing its own iron used in the making of farm implements and household items, including the well-known Franconia Stove. The 1889-built Delage Bridge is the oldest known iron bridge in town, as well as one of the oldest in the state. It is also important as one of just two lenticular pony truss bridges left in the state (see the next entry for the other), and one of four lenticular truss bridges of all types remaining in New Hampshire. This bridge was also unique in that it was essentially a private bridge, originally spanning the Ham Branch from Delage Farm Road to the home of Albert Delage. The bridge stood in its original location until 2001 when, deemed unsuitable for modern travel, it was lifted by crane from its original location and was relocated over a small creek emptying into the Gale River a short distance away on the grounds of the

The pinned joints of the Delage Bridge are easily visible where the upright members of the truss meet with the upper and lower chords.

town's visitor center. Here, one can view the bridge up close and personal and see all its construction details, as well as gain some insight into the town of Franconia's history from displays at the visitor center. Quite fittingly, just across the Gale River from this site sits an old iron furnace, the last such structure remaining in the state from days gone by.

20. Franconia — The Dow Avenue Bridge; DATE BUILT: ca. 1889; TRUSS TYPE: Lenticular pony; FEATURE CROSSED: Gale River; LENGTH: N/A; SPANS: 1; EASE OF ACCESS: Easy

Little is known of this historic lenticular truss bridge in Franconia, located just a short distance away from the Delage Bridge. Unlike most of the remaining truss bridges of this type, built by the Berlin Iron Bridge Company of Connecticut, this bridge no longer has its builder's plaque to identify the year it was built. That the Dow Avenue Bridge was built by the late 1880s is certain, as this bridge truss type fell out of favor by the early 1890s. The guess that this bridge was constructed about the year 1889 is based on the tendencies of the Berlin Iron Bridge Company and its salesmen. Good at their job in selling bridges, they often succeeded in their aggressive marketing approach to sell more than one bridge to a town. With the purchase of the Delage Bridge in 1889, it is not unreasonable to conclude that the Dow Avenue Bridge was contracted for by Franconia at about the same time, or within a year either way. The one-lane bridge today is in excellent condition, having been recently refurbished,

The Dow Avenue Bridge, Franconia, is the only lenticular truss bridge in all of northern New England that still carries highway traffic.

and even has new concrete abutments. It is further significant as the only lenticular truss bridge in all of northern New England that is still open to automobile traffic.

21. Bartlett-Hart's Location — Second, Third, and Fourth Iron Bridges; DATE BUILT: 1906; TRUSS TYPE: Baltimore; FEATURE CROSSED: Saco River and Sawyers River (Fourth Iron Bridge); LENGTH: 164 feet; SPANS: 1; EASE OF ACCESS: Moderate

These railroad bridges built by the Pennsylvania Steel Company of Steelton, Pennsylvania, are representative of a number of bridges built by the Maine Central Railroad on their Mountain Division, running from Portland, Maine, west through the mountains of New Hampshire and ending at St. Johnsbury, Vermont, in the 1906–1907 timeframe. This rail line was originally built by the Portland and Ogdensburg Railroad before being leased by the Maine Central in 1888. Passenger service continued, though gradually slowing down, into the 1950s; but the Mountain Division continued on with its freight business for many years until its successor, the Guilford Transportation Industries company, purchased the railroad in 1981 and discontinued the line two years later. The time period when these three identical "Iron" bridges, as well as four similar but slightly smaller bridges in neighboring Bartlett, were also built was a high point in the Mountain Division's prosperity.

Though fabricated of steel, their designation as an "Iron" bridge is a vestige of the past, when their predecessors were more lightly built spans truly made of iron. These newer bridges, now over 100 years old themselves, have held up very well over the years. Since the mid-

Top: The Second Iron Bridge, Bartlett. *Bottom:* The Third Iron Bridge, Bartlett. Note the builder's plaque on the right-side column.

1980s, the old Mountain Division line from Conway to Bretton Woods has been operated by the Conway Scenic Railroad, a heritage railroad founded in 1974 that runs both diesel locomotives and a steam locomotive. In season, daily maintenance runs are performed on the line and its bridges to make sure all conditions are suitable for operations. The area around Second Iron Bridge, as well the Third Iron Bridge, are easily accessible by a short hike, while the Fourth Iron Bridge is located near a popular hiking trail and can be easily seen from U.S. Route 302 in Hart's Location. One of the best ways to experience these bridges, however, is to take a ride on the Conway Scenic Railroad. No matter how you view them, all of these bridge sites, located within about fives miles of each other, are popular with hikers and are popular swimming areas in the summer season. Finally, if you're wondering whether or not there is a First Iron Bridge in addition to the bridges discussed above, if you guessed yes you would be correct; it crosses a small brook just out of North Conway and is not a truss bridge, but a small deck-plate girder bridge.

22. Hart's Location — Willey Brook Bridge; DATE BUILT: 1905; TRUSS TYPE: Warren deck; FEATURE CROSSED: Willey Brook; LENGTH: 400 feet; SPANS: 2; EASE OF ACCESS: Difficult

The Willey Brook Bridge, Hart's Location. Several diesel locomotives and passenger cars of the Conway Scenic Railroad making their way over the bridge (courtesy of Ryan Parent).

This impressive railroad bridge lies on the former tracks of the Mountain Division of the old Maine Central Railroad. Built in 1905, the bridge is now part of the operations of the Conway Scenic Railroad and is well cared for by both state bridge engineers and the CSR. Soaring 94 feet above Willey Brook, and spanning the chasm between Mount Willey and Mount Willard, the current bridge replaces an early webbed lattice truss bridge built in the 1870s. As if the scenery and its setting isn't reason enough to highlight the Willey Brook

Bridge, the area is also noted as being one of the most historic in the White Mountains. In the valley below this site was once the site of the Willey house, the home of Samuel Willey, Jr., his wife Polly, their five children, as well as two hired-hands. A massive rock and mud slide occurred in Crawford Notch on August 28, 1826, during a torrential rainstorm which made the Saco River rise 20 feet in a very short time and wiped out 21 wooden bridges in the valley. The path of the natural disaster went through the meadow where the Willey Family homestead was located, and when friends and relatives checked in on them several days later, they found the entire family gone. While a rock ledge above the house shielded the Willey home itself from the slide, the family hastily left the home to seek safety, leaving an open Bible on the table. They may have sought shelter in a rock enclosure further down from the house, or maybe they were making their way up the mountain to avoid the rising flood waters. Whatever their destination, the choice to leave their home was a fatal one, as all were killed in the slide. Willey Brook and Mount Willey were subsequently named after the family. For some years, until it burned down, the Willey house was open as a tourist attraction and the rock-strewn meadow where the house once stood retained an aura of wildness and even danger.

The area of the Willey Brook Bridge is equally interesting and also saw a bit of tragedy. Just beyond the western approach to the bridge for many years there was located a section house where railroad workers assigned to maintain this section of the Maine Central tracks were housed. This section house, established in 1887, became the home of Loring and Hattie Evans and their children in 1903, and here it was that their four children were raised. Even after Loring was killed in a train accident in 1916, Hattie stayed there on that isolated mountainside until 1942 and raised her children by herself. Today, if you should chance to hike to this area, you can see the cellar hole of the old house, torn down by the railroad in 1972, as well as an old boiler and other scraps of metal. Equally interesting is the graffiti etched on the rocks between the section house and the bridge dating back to the year 1875, near the time when the first bridge was built. The Willey Brook Bridge was recently repaired by the joint efforts of the state of New Hampshire and the Conway Scenic Railroad and is in excellent condition.

---

23. Hart's Location — Frankenstein Trestle; DATE BUILT: 1930; TRUSS TYPE: Steel Trestle; FEATURE CROSSED: Frankenstein Cliff; LENGTH: 600 feet; SPANS: 5; EASE OF ACCESS: Difficult

This bridge is one of the most storied attractions in New Hampshire's White Mountain region, and certainly one of the most historic. Like the Willey Brook Bridge, it lies on the original line built by the Portland and Ogdensburg Railroad in the late 1870s. Interestingly, one of the stipulations of the project by the state of New Hampshire was that the railroad line had to pass right through Crawford Notch, thus requiring some rather heroic bridges. In the late 1880s the line was leased and became the Mountain Division of the old Maine Central Railroad. The line is now used by the Conway Scenic Railroad after service ended in the early 1980s and the state bought the line. The original bridge at this site was constructed in the winter of 1874 to 1875 and was intended to be built of wood. However, the ship carrying its southern pine timbers was lost at sea and the P&O Railroad had to make other plans. This they did, contracting with the Niagara Bridge Works of Buffalo, New York to build the bridge of iron at a cost of 125 percent of the original planned wooden structure.

The bridge engineer was John Anderson, the brother of the president of the P&O Railroad,

The Frankenstein Trestle, Hart's Location.

Samuel Anderson. It was completed in June 1875 and measured 520 feet long and 85 feet high, bridging the chasm at Frankenstein Cliff. Though the bridge is a monster in size and height, it was not named after the famed literary creation of Mary Shelley, but artist Godfrey Frankenstein. Born in Germany, Frankenstein came to America at an early age and would gain renown as a painter in the 1850s, famed for his giant panoramic painting of Niagara Falls in the 1850s. He also visited the White Mountains and was friends with the largest landowner in the Crawford Notch area, Samuel Bemis, who had charged the railroad just one dollar for the use of his lands in the Notch. In fact, Frankenstein would later become a prominent member of the White Mountain School of painting, one closely allied with the Hudson River School in New York. He painted White Mountain scenery, including the cliff that would later bear his name. It was his friend and patron, the wealthy Samuel Bemis, who named the cliff and the bridge after Frankenstein, who died in 1873 before the bridge's construction had started.

The first trestle was a light and airy structure that nonetheless carried the loads required without difficulty. In 1892 the Frankenstein Trestle was rebuilt by the Union Bridge Company of Athens, Pennsylvania, of wrought iron as modern railroad equipment was getting bigger and heavier. The trestle was again worked on in 1930, getting steel bents and additional and stronger pier support, and was further strengthened again in 1950. Today, a hiking trail runs beneath the bridge, while the Conway Scenic Railroad runs excursions over the trestle. Either way, the view of the White Mountains from Frankenstein Trestle is a spectacular one.

24. Bretton Woods—Fabyan Station Bridge; DATE BUILT: 1892; TRUSS TYPE: Warren double-intersection; FEATURE CROSSED: Ammonoosuc River; LENGTH: N/A; SPANS: 1; EASE OF ACCESS: Easy

The Fabyan Station Bridge, Bretton Woods.

The railroad bridge at Fabyan is important not only for its truss design, a now rare double-intersection Warren truss, but also for the fact that it is the oldest metal truss railroad bridge in the state. Its builder was the Boston Bridge Works company, a prolific bridge builder in the region of both railroad and highway spans. The Fabyan Station and bridge just beyond are named after the famed grand hotel that once stood close by, the Fabyan House. This establishment was built in 1873 and owned by Sylvester Marsh, who built the nearby Mount Washington Cog Railway. The hotel, which replaced a previous one that had burned down twenty years prior, had accommodations for 500 people and was one of the prime tourist destinations in the White Mountains for many years before it burned in 1951. Very close to the bridge a spur line once led from Fabyan Station to the base of Mount Washington Cog Railway.

The original bridge on this site was a webbed lattice iron truss bridge, while another spur line crossed the Ammonoosuc River right next to it on a wooden trestle. Today, the Fabyan Station Bridge sees but limited seasonal use, as the Conway Scenic Railroad's 7470 steam locomotive and other trains make the journey from North Conway to Fabyan Station for only one month during the fall foliage season. However, the bridge and the tracks at Fabyan did experience some unusual activity in June 2009 when a 227-ton transformer for the utility company Public Service of New Hampshire was transported by the Conway Scenic Railroad using a specialized freight car. It was the first time since 1983 that any freight service had been experienced by the former Mountain Division line of the old Maine Central Railroad. To accommodate this high load, a portion of the overhead bracing at each end of the bridge had to be cut out, a common problem for older metal truss railroad bridges that are on active freight carrying lines.

---

25. Shelburne — Meadows Bridge; DATE BUILT: 1897; TRUSS TYPE: Pratt Combination (pin-connected); FEATURE CROSSED: Androscoggin River; LENGTH: 504 feet; SPANS: 4; EASE OF ACCESS: Moderate

The Meadows Bridge as it appeared in an early postcard view. Note the small approach span to the through-truss bridge, as well as the decorative cresting and builder's plaque above the portal.

Before it was bypassed in 1984 and a part of it was removed for safety reasons in 2004, the Meadows Bridge was once one of the grandest metal truss bridges in the state. It was built by the Groton Bridge Company of Groton, New York, in 1897 using a standard Pratt truss design with pin connections and is the largest pin-connected bridge still remaining in New Hampshire. The contract price for the bridge was $10,000, with the town of Shelburne originally responsible for $4,000 and the state paying $6,000. However, the legislature only approved $2,500 for the state share, thus leaving Shelburne to find the additional funds. This was achieved in a rather unusual manner when local citizens, both year-long residents and summer visitors, banded together to raise the funds privately. Indeed, the bridge was important to local tourism and provided access to two inns on the north side of the Androscoggin River. To get to these inns, carriages had to ford the river, but the Meadows Bridge changed all that and made access much easier and safer. One of the inns on the north side was the Philbrook Farm Inn, whose proprietor, Augustus Philbrook, was a town selectman and served in a voluntary capacity as the town's construction supervisor on the bridge project. The pride taken in the town's ability to erect such an important structure is well demonstrated by the decorative

The Meadows Bridge as it appears today, rusting and forlorn. As the years go by, hope is fading that this once majestic bridge will ever be restored. Photograph courtesy of Craig Hanchey.

cresting, builder's plaque, and star-topped finials above the end posts of the bridge. Flourishes such as these were very common on early iron bridges, but less so by the late 1890s.

The location of the Meadows Bridge at the fording place seems natural, but there had been discussion that the bridge might be constructed further upstream at a location known as Gates Crossing. However, a local resident, one Miss A. Whitney, did not want the bridge so close to her home and is said to have donated $1,000 to the bridge fund if it were moved to another location. She seems to have gotten her wish! The only other dispute, as it were, in the building of this bridge came from an adjacent property owner, Silas Morse, who received a settlement for property damages likely incurred when the bridge truss materials were staged on the riverbank adjacent to the area. The bridge, draped in three American flags, was dedicated for service in October 1897. One of the grandest celebrations the town had ever seen took place; a procession of more than one hundred teams, led by a brass band from nearby Gorham, crossed and re-crossed the bridge, after which speeches were given and a public prayer was offered. The bridge served well for nearly 90 years before its closure in 1984 when a new bridge was built. For years the Meadows Bridge served as a unique pedestrian walkway, but in 2004 this ended when it was decided to remove two spans of the bridge to the riverbank because its unique pier system was failing because of the changing course of the river. Today, the future of this bridge is uncertain doubt due to a lack of funding options. It now sits rusting away, waiting for its fate to be determined. Among the options discussed have been the repair of the piers and re-erection of the bridge in its original location, or possibly erecting one of the spans elsewhere in town.

12

Maine Bridges

1. Biddeford — Elm Street Bridge; DATE BUILT: 1929; TRUSS TYPE: Baltimore; FEATURE CROSSED: Elm Street; LENGTH: 139 feet; SPANS: 1; EASE OF ACCESS: Moderate

The Elm Street Bridge, Biddeford.

This heavily skewed bridge was fabricated by the Phoenix Bridge Company of Phoenixville, Pennsylvania, and is important as a representative example of the strongly built railroad bridges of the day suitable for the heavier railroad locomotives and rolling stock that were used in the first half of the 1900s. This dual track bridge has three trusses and is located on an old Boston and Maine Railroad line as it ran north to Portland and south to South Berwick and beyond; one of the sets of tracks are abandoned, while the other set is still active. Near the site of the Elm Street Bridge are several abandoned factory buildings dating from the early 1900s, as well as the old Biddeford passenger station just to the south.

---

2. Portland — St. John Street Underpass; DATE BUILT: 1890; TRUSS TYPE: Baltimore; FEATURE CROSSED: St. John Street; LENGTH: 117 feet; SPANS: 1; EASE OF ACCESS: Easy

The St. John Street Underpass is Maine's oldest surviving metal truss railroad bridge and, located in this urban setting, surely one of the most visible. While the bridge utilizes the

The St. John Street Underpass, Portland. The oldest metal truss bridge in the state, this bridge still carries one lane of active railroad traffic.

Baltimore truss, one common amongst railroad bridges, the use of riveted connections at a time when pin-connections were more common speaks to the extra investment the railroad was willing to make on this important structure to ensure that it could handle the heavy loads that rolled its way. Because of the heavy rail traffic in this important port city, the bridge once had a dual set of tracks, and has three trusses. Today, one of these sets of tracks has been abandoned and the bridge is no longer decked on one side. The bridge was once part of the Maine Central Railroad line, and their extensive yards are located just to the south of the bridge, close-by the system of tracks at the Portland Terminal. The St. John Street Underpass was built by the Boston Bridgeworks company, established in 1879 by D.H. Andrews and one of the most prolific railroad bridge fabricators in New England for many years. This bridge is the oldest known example in northern New England of one of their riveted bridges.

---

3. Windham — Gambo Falls Bridge; DATE BUILT: 1912; TRUSS TYPE: Warren pony; FEATURE CROSSED: Presumpscot River; LENGTH: 77 feet; SPANS: 1; EASE OF ACCESS: Easy

The setting of this bridge in a quiet wooded area is in sharp contrast to its early years. The Gambo Falls area along the Presumpscot River between Windham and Gorham was once the site of one of America's largest gunpowder mills. The Oriental Powder Company operated here for 71 years beginning in 1834 and was so large that it supplied 25 percent of the gunpowder used by the Union Army during the Civil War. For about a mile along each side of the river at Gambo Falls the area was filled with mill buildings, refineries, and storehouses

The newly restored Gambo Falls Bridge connects Windham and Gorham, across the Presumpscot River.

and was a beehive of activity. The buildings in the powder mill were widely spaced out so as to minimize any damage in case of an accidental explosion at one of the facilities. After the Oriental Powder Company went out of business, the mills were at one time operated by the giant chemical company Dupont, and it was under their management that the current Gambo Falls Bridge was built in 1912. Once the Dupont operations ceased, the old factory was abandoned and was eventually dismantled over the years, though some remnants are visible in the vicinity of the bridge. Indeed, the Gambo Falls Bridge is one of the last major vestiges of the mills along the Presumpscot River at Gambo Falls, and is important as the only metal truss bridge remaining in Maine with direct factory associations. Just as the bridge languished for many years, so did the Presumpscot River. The river, which runs from Sebago Lake south to its outlet at Casco Bay, about 25 miles, was once one of the most heavily dammed rivers in America, and one of the most polluted. Of the nine dams that remain on the river, including one at Gambo Falls, all are owned by a South African Company. The Gambo Falls Bridge after lying abandoned for many years, was restored in 2005 and is part of a recreational trail connecting Windham on one side of the river with Gorham on the other.

---

4. Brunswick-Topsham — The Swinging Bridge; DATE BUILT: 1892; TRUSS TYPE: Suspension; FEATURE CROSSED: Androscoggin River; LENGTH: 332 feet; SPANS: 1; EASE OF ACCESS: Easy

This attractive bridge has been recently renovated and is among a small collection of historic pedestrian suspension bridges remaining in Maine. Though it appears purely recreational today, it was actually a private factory bridge when first built, allowing those living in the

The Swinging Bridge between Brunswick and Topsham.

Topsham Heights area to cross the river, where many worked at the Cabot Cotton Mill in Brunswick. Indeed, Cabot owned land right to the bridge site, but granted an easement for those crossing over their land. Though the developers hoped that the bridge would increase the value of house lots in Topsham Heights, this did not happen and the bridge was a losing proposition for them. By 1906 both Topsham and Brunswick took control of the bridge as a public highway. The bridge was completed in the fall of 1892 and was built by the John Roebling Sons Company of Trenton, New Jersey, the same family that pioneered the use of wire cable in American bridges and built the Brooklyn Bridge. The towers of the bridge were originally built of wood, but were replaced with steel towers in 1916, manufactured by Megquier and Jones of South Portland, Maine.

Though the bridge has served well over the years, the Flood of 1936 nearly killed it. Luckily, while the deck of the bridge was wrecked, the towers and cables survived and the bridge was rebuilt with the help of the WPA. The bridge remained in operation thereafter for many years, but with very little maintenance; its towers becoming very rusty, its cables corroded, and its deck unsafe for more than a small group of people. Thankfully, the town banded together to preserve the bridge, establishing a Save Our Swinging Bridge group dedicated to publicizing the plight of the bridge and leading the way in fundraising efforts. In 2008 the bridge was finally restored at an estimated cost of over $600,000, including about $400,000 from the federal government, $100,000 from the Maine DOT, and the rest from private donations. Today, the bridge has regained its place within the towns of Brunswick and Topsham as a destination for both local pedestrians and curious tourists alike.

---

5. Brunswick-Topsham — Free Black Bridge; DATE BUILT: 1909; TRUSS TYPE: Baltimore with suspended deck; FEATURE CROSSED: Androscoggin River; LENGTH: 318 feet; SPANS: 2; EASE OF ACCESS: Easy

The Free Black Bridge, Brunswick-Topsham.

The Brunswick-Topsham area is surpassed by few other towns in northern New England for its number of historic metal bridges. The three-span, 815 foot long Frank J. Wood Bridge further upstream is the longest in Brunswick, while the Swinging Bridge (see previous entry above) is the oldest and of a rarer type. Despite this, the Free Black Bridge is by far the most interesting, and the most exciting of all the three bridges to cross. The Free Black Bridge, a combination bridge serving both railroad and highway traffic, was built by the Maine Central Railroad in 1909 to replace a previous combination bridge. The bridge utilizes the heavy Baltimore truss design then in common use by New England railroads and was fabricated by the Pennsylvania Steel Company of Steelton, Pennsylvania. If it were just a railroad bridge, the Free Black Bridge would not be a rarity, but just one of many Baltimore truss bridge built by the MCR in the early 1900s as it was updating its bridge inventory from earlier iron bridges to heavier capacity steel spans. What makes this structure so unusual is its suspended lower deck for automobile traffic. The deck is suspended from the lower chord of the bridge trusses above by metal eye bars that are pin-connected at each panel point on the truss. The suspended deck itself consists of metal floor beams with timber stringers supporting a wooden plank deck. This ingenious solution for adapting the bridge to two modes of travel is not only interesting, but rare. No other metal truss bridge like the Free Black Bridge is found in northern New England, and possibly all of New England.

This bridge is an experience to cross, as the suspended roadway is very narrow, about twelve feet, while the overhead clearance is just short of eleven feet. Because the roadway is a single lane, vehicles taking this route have to be watchful of oncoming traffic at the approaches at each end and the experience is very similar to traveling through a covered bridge or a well-lit tunnel. The approach located on the Topsham side is additionally unique, and harrowing, as it involves a sharp angled curve. Despite this dangerous approach, locals are used to it and few accidents seem to occur. The rail line served by the upper portion of the Free Black Bridge was once a part of the Androscoggin Railroad, which ran from Brunswick to Leeds when it was first established, and was later leased by the Maine Central in the 1870s before they eventually purchased it in 1911. It was later known as the Lower Road of the MCR, running from Brunswick to Augusta. The rail line is still active today but has limited use for local traffic.

---

6. Durham-Lisbon — Durham-Lisbon Bridge; DATE BUILT: 1937; TRUSS TYPE: Continuous Warren; FEATURE CROSSED: Androscoggin River; LENGTH: 363 feet; SPANS: 2; EASE OF ACCESS: Easy

The south span of the Durham-Lisbon Bridge.

This bridge is one of several dozen bridges built by the Maine State Highway Commission in the aftermath of the floods in March 1936, replacing an earlier and more lightly built iron bridge. In fact, the bridge's current pier previously served the older iron bridge and survived the 1936 flood while the old bridge was swept away. For sometime afterward, workers at the mill coming from Durham had to use an improvised ferry system — Earnest Learnards 27-foot ocean-going (usually!) lobster boat — to get across the river, and later a temporary wooden bridge was built. Of the many flood-replacement bridges built in Maine, three, including the Durham Bridge, were continuous Warren truss bridges. This type of truss bridge was built in increasingly greater numbers for long-span structures during the 1930s, once their capabilities were fully understood as the science of stress analysis made great advances. Continuous truss bridges were also especially important during the Depression Era, when economy in bridge building was a must. The Durham Bridge is one of the state's earliest-built bridges of this type, which later became more common after World War II.

Of course, the rebuilding of this bridge, like so many others nationwide during the 1930s, would not have been possible without funding from President Roosevelt's New Deal administration and its many "alphabet" agencies, in this case the PWA (Public Works Administration). The Durham Bridge is also significant as it served as an access point for the nearby Worumbo Company mill (and several others) in Lisbon. Founded in 1864, the company eventually became quite prosperous and saw a great period of expansion in the 1920s and 1930s, but it eventually closed in 1964. Today the surviving mill building, a giant white structure, is on the National Historic Register and employs a small amount of people at a factory outlet store for blankets and other textile items. A visit to the Durham Bridge in person and a stroll on its pedestrian sidewalk also hints at the span's former importance to the many mill workers that lived across the river in Durham. The remaining decorative light fixtures, almost all of which are now inoperable, remind us that these workers crossed the bridge in the early hours before daybreak and yet again when returning home after night had fallen.

---

7. Bath — Carlton Bridge; DATE BUILT: 1926; TRUSS TYPE: Warren truss vertical lift; FEATURE CROSSED: Kennebec River; LENGTH: 3,098 feet; SPANS: 7; EASE OF ACCESS: Easy

This distinctive bridge, which has been deemed to be a non-historic span by the state, is the second of three vertical lift bridges to have been built in Maine from 1920 to 1940, and the only one entirely within the state (the other two are interstate bridges shared with New Hampshire). It was built by the Kennebec Bridge Commission and served as a toll bridge for many years before becoming a free crossing. The Carlton Bridge, named after state senator Frank Carlton of Woolwich who chaired the bridge commission, was designed by J.A.L. Waddell and was fabricated by the steel giant McClintic-Marshall, an arm of Bethlehem Steel. The bridge was built not only for highway traffic above, but also the tracks of the Maine Central Railroad beneath, the crossing at this location having been previously served by a ferry which carried railroad cars across the river to Woolwich. The construction of the Carlton Bridge's pier supports and foundation was quite a feat of engineering at the time, using caissons sunk to record depths and quick hardening cement. Its spans consist of a 234-foot long vertical lift span and six Warren deck approach spans, four to the east of the vertical lift span and two to the west.

With the aging of the bridge over the years and the downtown congestion in Bath resulting from its two lanes and heavy traffic (about 25,000 cars daily), the discussion to replace the

The Carlton Bridge, Bath.

bridge began in the 1980s. However, it was not until the 1990s that plans to build a new high rise concrete bridge as a replacement proceeded apace. This new span, the Sagadahoc Bridge, was built from 1998 to 2000, resulting in the closure of the Carlton Bridge to highway traffic upon its completion. While the two eastern-most spans of the Carlton Bridge have been removed, leaving it with an odd truncated appearance at the Bath end, the bridge still carries Maine Eastern Railroad freight and passenger traffic over the Kennebec on its lower deck. Though closed to highway traffic for over ten years now, the vertical lift span still operates to allow the passage of commercial maritime traffic. The bridge is still important for the railroad traffic it carries over the Kennebec, something the four-lane Sagadahoc Bridge was not designed for. Though the Carlton Bridge is a later version of Waddell's vertical lift bridge and is thus not historic in this regard, it is notable for having brought the era of the automobile to Bath and the mid–Maine coastal region. In its first year alone, the Carlton Bridge served one million cars, and, as a result, the first traffic light in the city of Bath was soon installed at an intersection near the bridge. Carlton Bridge's high-rise profile and its location near the Bath Iron Works Shipyard have made it a prominent feature on the local skyline for many years now, but for how much longer it will survive is unknown. As of late 2010 there were no plans by the Maine DOT to replace the bridge.

8. Arrowsic — Max L. Wilder Memorial Bridge; Date Built: 1950; Truss Type: Cantilever arch; Feature Crossed: Sassanoa River; Length: 838 feet; Spans: 3; Ease of Access: Moderate

The Max. L. Wilder Memorial Bridge, Arrowsic.

This bridge is important as one of only two cantilever truss bridges ever built in the state of Maine. Formerly known as the Sasanoa River Bridge, the Max L. Wilder Memorial Bridge was renamed for the state bridge division chief in charge of building the bridge after his death in 1962. The bridge is important to the local area, carrying Route 127 over the Sassanoa River (part of the Kennebec River estuary) connecting Arrowsic and Georgetown islands with Woolwich. The decision to build a cantilever bridge at this location was based on its difficult location and steep approach, as well the long length required. It was built with little falsework, as is typical with bridges of this type, its anchor arms supporting deck trusses spanning from the shore to the piers on either side and transitioning into through truss cantilever arms over the river's main channel. The Max L. Wilder Bridge is the newest bridge included in this study and is well maintained by the state, expecting to undergo an $800,000 resurfacing project in 2011.

9. Southport — Southport Bridge or Townsend Gut Bridge; Date Built: 1939; Truss Type: Warren truss swing span; Feature Crossed: Townsend Gut; Length: 374 feet; Spans: 2; Ease of Access: Easy

If you're timing is correct, this might just be one of the most fun, not to mention interesting, of all the bridges you can visit in Maine. In fact, it is the only historic moveable bridge

The Southport Bridge, Southport, crosses Townsend Gut.

entirely within the state that is manned night and day, every day of the year, and therein lies its importance. The bridge design itself was a standard one used by the state of Maine and they built eight bridges of this type from 1901 to 1954. The previous bridge at this location was a wooden moveable bridge that was also a toll bridge; the walk to church on Sundays was free, but at other times pedestrians had to pay 5 cents to cross, while bikers were charged 25 cents and a horse and buggy cost $2.50. The new bridge does not appear to have been a toll bridge. While this bridge is not only the only one that is manned today it is one of the largest, surpassed in size only by the Kennebec River Bridge in Richmond (see below). Because of the slow nature of the operation of the swing bridge type of moveable spans, most have been replaced with bridges that are fixed in place and higher over the water to allow uninterrupted traffic flow on land and water. The bridge was fabricated by the Lackawanna Steel Construction Company, but local tradition states that it was hauled to the bridge site on a barge captained by Mace Carter, a local mariner. After all, who would know the waters hereabouts better than a local sea captain!

The Townsend Gut Bridge has remained in operation for over 70 years now because it remains suitable for the local traffic; though heavy in the summer tourism season, it does not require a wide roadway, nor does the bridge's pivot pier obstruct navigation, which largely consists of small pleasure craft and fishing boats. Though early swing bridges in America were operated by hand, this bridge, like all modern swing bridges, is now operated by electric motors with reduction gears and shafts to the pinion gear located on the pivot pier, with the bridge movement guided by balance wheels traveling on a track. Look over the side of the bridge when it is fixed in position and you can see the timber piling that acts as a fender for the bridge, while above the center of the bridge is a small building that is not an operator's house, but actually houses the bridge's electrical machinery. On the northwestern corner of the bridge is the operator's house, which is cantilevered out over the waterway and was added in 1973. Inside the house, which also provides shelter during inclement weather, is a small cot and working area and kitchen, as well as a phone. The gates to block traffic when the bridge is in operation were originally manually operated and still remain in use, though automatic gates have also been added.

By far the most important, and interesting, feature of this bridge is its human components; for the majority of the bridge's life, it has been manned by a member of the Lewis family. Currently, the bridge is primarily manned by twin brothers, Duane and Dwight Lewis. I had the pleasure of meeting Duane while he was manning his post in April 2010 and found him a treasure trove of knowledge. It was his father Norman Lewis that was the first in the family to man the bridge, doing so for 42 years. He was succeeded as chief operator by his sons upon his retirement, and a plaque was placed on the bridge in honor of his service in 2002. Of course, Duane and Dwight had already worked on the bridge for a number of years with their father and were familiar with the intricacies of the bridge operations. Regarding these operations, Duane's comments are insightful:

Bridge operator Duane Lewis of the Southport Bridge. The Lewis family members have been the operators and caretakers for this important bridge during its entire existence.

> We generally work 8 hour shifts on a rotating basis, 4 p.m. to midnight, and so on, and I switch off with my twin brother Dwight. The bridge opens in the off season maybe once per shift, but during the summer season 10–15 times a day. The bridge is greased three times a week. I'm called "Grease Monkey Lou" and I love that part of the job. Generally, one man greases the bridge, another man pivots the bridge. There's no special equipment other than a grease gun and coveralls. You've got to be an agile climber underneath the bridge! How do you know when the bridge needs greased? When the bridge makes a clanking noise! As far as accidents go, they're rare...one fellow ignored the warning lights and cracked his windshield when he slammed into the gate. The bridge takes about five minutes to pivot fully open, and another five to close. Motorists, even tourists, are generally patient...boat accidents are rare, though sometimes they go through the wrong channel. We get some rough weather sometimes, but the bridge does fine. I can only remember the bridge being closed a few times, once when there were 40 knot winds and we had to chain both ends of the bridge. Fishing is allowed off parts of the bridge with a salt-water license, and sometimes we

even fish. This bridge is in good shape and will last another 50 years if it is taken care of (Lewis Interview, 4/24/2010).

Interestingly, it is not just Norman Lewis (42 years) and his twin sons, Duane (46 years) and Dwight (46 years) who have worked on the bridge, but also two other brothers, Donald (10 years) and Roy (1 year) and even their sister, Ruth Lewis Snowman (10 years) who, according to Duane, did the job as well as any of them did. The Lewis family, which has over 150 years of combined service attending the bridge now, used to live in a house by the rocks at the south end of the bridge, so the trip to work was not a long one. Duane and Dwight's mother, Mildred Lewis, sometimes brought her husband and her sons their lunch or dinner and taught her sons how to cook for themselves. Later the family moved to a new homestead about a mile from the bridge, where Duane still lives. Today, Duane Lewis is the best bridge ambassador the state of Maine has. Not only does he give local lectures about the bridge and its history, but he also loves to talk with pedestrians walking over the bridge or anyone else who might happen to stop on by.

---

10. Richmond — Kennebec River Bridge; DATE BUILT: 1931; TRUSS TYPE: Parker truss swing span; FEATURE CROSSED: Kennebec River; LENGTH: 1,239 feet; SPANS: 10; EASE OF ACCESS: Moderate

The Kennebec River Bridge, Richmond.

The Kennebec River Bridge, quite simply, is one of the most imposing bridges in the state. It is the longest of all the moveable bridges remaining entirely within the state of Maine, as well as one of the longest metal truss bridges overall, spanning the broad Kennebec River midway between the port city of Bath and the capital city of Augusta. The city of Richmond was once a renowned shipbuilding center, but by the time this bridge was built its maritime prominence had largely vanished and now it is a popular tourist town. The Kennebec River Bridge has ten spans overall, including five small steel stringer approach spans, four Parker truss spans, and a center Parker truss swing span measuring 177 feet long. Originally completed in 1931, the bridge was severely damaged by the 1936 floods, losing the swing span and the two western truss spans, all of which were replaced in 1937 using the original builder's plans. The bridge was constructed by the American Bridge Company and has proven to be well-built, surviving the later floods on the mighty Kennebec.

The bridge was once manned year round, but is now only operational from June through September. With the greatly reduced maritime traffic on this part of the river, the only moveable action the bridge sees during the cold winter months is the occasional passage of a Coast Guard cutter on ice-breaking duties. Despite the fact of its only occasional use as a swing span, the bridge is well maintained and is an important crossing. In 1992 the bridge's electrical system and center bearings were replaced. Previously, the original deck was replaced in 1959 with an open steel grid deck and the timber pile fenders were replaced in 1988 with ones of steel and concrete. The Kennebec River Bridge has two operator houses located on the western span adjacent to the swing span, manually operated traffic gates, and a house above the swing span that houses the electric motor that supplies the power to pivot the bridge. The bridge was originally operated by the directors of the Maine Kennebec Bridge Company and was a toll bridge until August 1949. The funds raised by these tolls help offset the price of this long bridge, as well as others built during the same time within Maine. Though somewhat rusted today, the bridge is in sound structural condition and should see many more years of service.

11. Wiscasset — Sheepscot River Railroad Bridge; Date Built: 1893; Truss Type: Combination Warren and Pennsylvania; Feature Crossed: Sheepscot River; Length: 612 feet; Spans: 4; Ease of Access: Moderate

When this bridge was built in 1893 by the Maine Central Railroad, it was one of the most heavily built railroad bridges in the state, and one of the first to be built of steel. It measures just over 600 feet long and one of its spans was originally a small rolling bascule bridge, measuring somewhere near 60 feet long. The bascule bridge, which allowed access for small fishing and pleasure craft, was fixed in place many years ago and is now inoperable. Small moveable spans were once common on Maine's coastal railroad bridges so as not to impede the numerous small fishing craft and merchant cargo carriers that sailed these Downeast waters. The bridge lies on the Rockland Branch of the Maine Central and saw passenger and freight service for many years. However, passenger service was discontinued by the Maine Central in 1959 and the Rockland Branch, and its bridges, was acquired by the state when the line was in danger of becoming abandoned in 1987. Today, the Maine Eastern Railroad operates the line to Rockland and a ride on one of their passenger trains is one of the best ways to view the Sheepscot River Railroad Bridge. This impressive crossing, which is also visible from the U.S. Route 1 highway bridge over the river just to the south, is a rare example of both a Pennsylvania truss railroad bridge and an early pin-connected span. The bridge is well cared for and received a major overhaul in 2000.

---

12. Thomaston — Wadsworth Street Bridge; DATE BUILT: 1928; TRUSS TYPE: Pratt/Pennsylvania, former bascule bridge; FEATURE CROSSED: St. George River; LENGTH: 231 feet; SPANS: 2; EASE OF ACCESS: Easy

Opposite page: This ca. 1900 postcard view shows the Sheepscot River bridge with its original moveable span at far right, which is now a fixed span. *Top:* The Wadsworth Street Bridge, Thomaston (courtesy of Craig Hanchey).

The choice of bridge type made by the state of Maine when it came to the building of this distinctive span is somewhat of a mystery. By the time it was fabricated by the Boston Bridge Works in 1928, this rolling counterweight bascule bridge was considered an obsolete design and one that was noted for being difficult to maintain, as most bascule bridges used fixed counterweights by this time. The Wadsworth Street Bridge is now a true rarity, it being the only surviving example of this type of moveable bridge remaining in Maine, as well as the state's only historic bascule bridge overall.

Though no longer operational, the bridge is also the only historic bascule bridge in all of northern New England. Another odd thing about this bridge is the fact that though the Maine State Highway Commission's own bridge division began work on designing the structure in 1925, it was not completed until three years later. Was their concern and debate within the bridge division about whether or not this was the best type of bridge for this site? Either way, it never really mattered; by the time the Wadsworth Street Bridge was built, maritime trade on the St. George River was in steep decline. The bridge remained an operational bascule span for about the first twenty years of its life, but saw only sporadic occasions for use by 1950. About 1966, probably never having opened in years, the bridge was made a fixed span. Though the curved Pennsylvania truss portion of the bridge (72 feet in length) that acted as the counterweight tower and track still remains, all the other operating mechanisms, such as motors and gears, counterweights, chain lifts and wheels, and the operator's house, were removed for good. The lift span itself measures 55 feet in length and is a Pratt pony truss. Though now fixed in place, this span does retain one key component, its trunnions, that point to its earlier use as a moveable bridge. The north span on the Wadsworth Street Bridge is a 100-foot long Pratt truss span and is in itself significant as the oldest riveted Pratt truss highway span in the state. Just how rare the Wadsworth Street Bridge is may be gauged by the fact that there is only one known operational rolling counterweight bascule bridge remaining in the country at this time, the 1898-built Glimmer Glass Bridge in Manasquan, New Jersey. How many of this bridge type remaining nationwide that, like the Wadsworth Street Bridge, have retained their form is unknown but likely few in number.

---

13. Deer Isle — Deer Isle–Sedgewick Bridge; DATE BUILT: 1939; TRUSS TYPE: Suspension; FEATURE CROSSED: Eggemoggin Reach; LENGTH: 2,308 feet; SPANS: 3; EASE OF ACCESS: Moderate

Believe it or not, this bridge was considered by many at its time of building as an ordinary, run of the mill suspension bridge, inexpensively built. Even its designers, David B. Steinman and Holton Robinson, one of the premier bridge designing teams in the entire country, were seemingly less than enthusiastic. Steinman would later say of this bridge, "The P.W.A. gave us only one week to design the bridge. I needed three million dollars, but all they allowed me was $1,300,000. They said, 'Give us a bridge that will just stand up.' That's what it was. The hardships of the Depression affected not only the finances of engineers, but also their designs and, in the case of the Deer Isle and Thousand Islands Bridges, I was no exception to the general rule" (Ratigan, pgs. 244–25).

While the Deer Isle Bridge may have been considered the poor step-sister, perhaps, to the Waldo-Hancock Bridge, an award winning suspension bridge also designed by Robinson and Steinman, it is today Maine's longest operating clear span bridge, its main span measuring 1,088 feet in length. The bridge replaced an inadequate ferry system between Deer Isle and Sedgewick on the mainland and was important in opening up tourism to the area by carrying two lanes of State Route 15 over Eggemoggin Reach. Because of the unusual nature of the bridge site, a number of challenges had to be met by the bridge's designers, Steinman and Robinson, as well as the bridge fabricator, the Phoenix Bridge Company (Phoenixville, Pennsylvania), and the contractor involved in constructing the bridge, Merritt-Chapman & Scott (New York). One major concern was the height of the bridge over Eggemoggin Reach, a popular boating area for yachts and large sailboats. To accommodate such craft, the designers resorted to a steep approach grade and a vertical curve at the center of the main span, allowing for a clearance above water of 85 feet. Also posing a challenge was the construction of the main tower pedestals, which were built using prefabricated steel dams that were cut to fit the irregular bedrock below water after layers of mud had been removed and then filled with concrete. The cabling system used on the bridge was also fairly new, and one that became a cause for concern within a few

The Deer Isle Sedgewick Bridge, Deer Isle. This notable Depression era bridge was featured on the Maine State Highway map for 1939–1940 (courtesy of Maine DOT).

years. It was developed by Holton Robinson and consisted of prestressed twisted strand cables, first used on Steinman and Robinson's Waldo-Hancock Bridge in Maine and the St. John's Bridge in Portland, Oregon back in 1931. Each main strand of cables was connected to an anchorage rod by sleeve nuts that could be periodically adjusted. Even before the bridge was completed, heavy winds caused the light deck of the Deer Isle-Sedgewick Bridge to oscillate, prompting the addition of diagonal cable stays from the main cables to the stiffening girders on each of the bridge's towers. This was a lesson learned by Steinman during the final construction phase of the Thousand Islands International Bridge between the U.S. and Canada in 1938, when he first learned of this potential wind problem.

All suspension bridges in the U.S. became suspect in 1940, when the Deer Isle-Sedgewick Bridge was just a year old, as a result of the collapse of the Tacoma Narrows Bridge (nicknamed "Galloping Gertie") in Washington state within four months of its completion due to high winds that caused the bridge's deck to oscillate wildly. However, Steinman and Robinson had already come up with a remedy for their bridges.

Indeed, David B. Steinman would later admit that his Deer Isle bridge might have been even more at risk than the Tacoma Narrows Bridge had he not employed his system of cable stays as a preventative. However, even more and stronger longitudinal and transverse diagonal stays were added to the bridge after heavy bridge movement during the severe winter of 1942–1943 caused great damage and destroyed some of the stays. Yet another interesting aspect of this bridge's construction was its required completion date in the early summer of 1939. Because of this, the bridge work had to be done during the winter and early spring months, when severe weather abounds in northern New England. However, the use of prefabricated parts and working quickly between periods of high tides resulted in a bridge that was built on time and under budget. Today, the Deer Isle- Sedgewick Bridge is one of the most impressive and majestic spans in the state and is well maintained, having undergone a resurfacing project and receiving a new coat of paint in 2009–2010.

14. Prospect-Verona — Waldo-Hancock Bridge*; DATE BUILT: 1931; TRUSS TYPE: Suspension; FEATURE CROSSED: Penobscot River; LENGTH: 2,040 feet; SPANS: 3; EASE OF ACCESS: Moderate

The wild beauty of the Penobscot River setting of the Waldo-Hancock Bridge is evident in this postcard view.

The pending fate of this significant bridge, for so long a fixture on the mid–Maine coastal transportation network, will mark the sad end of a truly revolutionary and storied span. The

Waldo-Hancock Bridge, named for the two Maine counties which it connects, has now been closed since 2007 after a replacement bridge was built nearby over the Penobscot River. It is slated for demolition sometime in 2011–2012. Today the bridge sits sad and forlorn, its roadway closed at both ends, its towers rusting, and its steel cables now too badly corroded to safely carry vehicular traffic.

However, in its time the Waldo-Hancock Bridge was an important structure and one noted for its firsts. The bridge carried U.S. Route 1 over the Penobscot River and was the last link, so to speak, in Maine's coastal transportation network, the last of the big rivers to gain a crossing. Prior to the completion of the bridge, motorists either had to cross via a ferry service that was at times unreliable, or travel to cross the Penobscot River at Bangor some fifteen miles to the north. The Waldo-Hancock Bridge was designed by those famed suspension bridge builders, David B. Steinman and Holton Robinson, and was noteworthy for a variety of innovations. It was one of the first bridges of its kind in America to use prestressed and prefabricated wire cables, each of which consisted of 1,369 individual wires, to support the bridge. The Waldo-Hancock Bridge is also noted as being the first bridge in America to employ upright Vierendeel trusses for its two towers, a design invented by a Belgian civil engineer, Arthur Vierendeel. This truss, which was later used on New York's Triborough Bridge and the Golden Gate Bridge in San Francisco, is distinctive because of the fact that it has no diagonal members, but instead has rectangular openings, and was characterized by Steinman as being "a new artistic type emphasizing horizontal and vertical lines" (Steinman and Watson, pg. 330). Steinman was not only noted for his technical bridge designing skills, but also for his ability to design the perfect bridge for a given location. In regards to the Waldo-Hancock Bridge, its rocky setting on the Penobscot River, with old Fort Knox and colonial houses nearby "called for simplicity of design. An attempt, therefore, was made to secure an artistic effect with straight lines, by spacing, arrangement, and proportion" (Steinman and Watson, pg. 331). This, indeed, is just what Steinman did, and though the bridge was not the largest or most celebrated of the many heroic suspension bridges he built both nationally and worldwide, it may have been one of his favorites. One biographer of Steinman states that "he always felt that, even if the suspension bridge in Maine were his only achievement, his life would have been well justified" (Ratigan, pg. 208). The construction of the Waldo-Hancock Bridge proceeded apace without incident, in and of itself quite an achievement for any bridge project. Fabricated by American Bridge, it was Maine's first long span bridge and was the first bridge to cross the Penobscot River below Bangor. Interestingly, one of the young civil engineers on Steinman's New York design staff was Ray Boynton, a native of Skowhegan, Maine, and he took great pride in being assigned to work on this notable bridge in his home state. The resident engineer for the project was another of Steinman's staffers, Carl Gronquist.

The Waldo-Hancock Bridge, which sits 135 feet above the water, did pose a number of challenges to the contractor Merritt-Chapman & Scott, in charge of building the substructure, including battling the Maine winter in general and the construction of its foundations on an irregular rocky slope. To enable the efficient cabling of the bridge, a number of temporary footbridges were erected to aid in the process, while the cables themselves were marked and formed in advance, a technique never before attempted but now commonly used. In the end, David Steinman built a bridge that Maine could be proud of in every way. Completed and open for traffic on November 16, 1931, after just sixteen months of construction, the bridge cost only $850,000, well under the $1.2 million appropriated for the project. In fact, with this significant savings, the state of Maine, in a true "two-fer" deal, commissioned Steinman to build a smaller bridge, the Verona Island Bridge, between Verona and Bucksport and still had money leftover!

The newly built Waldo-Hancock Bridge gained national recognition soon after it was built when it received the annual merit award of the American Institute of Steel Construction as the Most Beautiful Steel Bridge built in 1931. The Waldo-Hancock Bridge was originally operated as a toll bridge before it became free in 1953. With early toll rates in the 1930s set at .55 cents for a passenger car, the price to use the bridge was not cheap, but one that motorists were more than willing to pay. Standing and serving as a stunning regional landmark for many years, the Waldo-Hancock Bridge was added to the National Register of Historic Places in 1985, but even this notable status could not save the bridge. Problems with the bridge surfaced in 1992 when a detailed inspection of one of its cables revealed a number of broken wires due to corrosion from water infiltration. Further inspections revealed even more problems with the cables, but it was not until eight years later, in 2000, that work was begun on strengthening the north cable, followed by work on the south cable beginning in 2003 with a contracted price of $4 million. Despite these efforts, the Waldo-Hancock Bridge was deemed to be too far gone, and a new span, the Penobscot Narrows Bridge, was built alongside it, opening in December 2006. Though effectively abandoned as of this writing and awaiting the day when it will be torn down, the Waldo-Hancock Bridge still retains a bit of its old glory when viewed at the right time of day. One can only hope that some portion of the bridge, perhaps a part of one of its Vierendeel truss towers, will be saved for posterity.

---

15. Waterville — Two Cent Bridge; DATE BUILT: 1903; TRUSS TYPE: Suspension with Warren pony approach; FEATURE CROSSED: Kennebec River; LENGTH: 400 feet; SPANS: 2; EASE OF ACCESS: Easy

The Two-Cent Bridge, Waterville.

The state of Maine is notable for its many surviving factory footbridges, and this span is one of the most interesting of them all. Like many such bridges, this structure was built as a private, money-making venture, or so it was hoped, to allow for the convenient passage of workers living in Waterville to paper factories across the Kennebec River in Winslow. It was owned by the Ticonic Footbridge Company, named after the Ticonic Falls located just downstream, and was designed by Edwin Graves. Born in 1865 and a native of Orono, Maine, Graves was a civil engineering graduate at the University of Maine and later moved to Connecticut by the 1890s. Here he was active as a prominent bridge builder, designing several Connecticut River bridges at Thompsonville and Middletown, as well as one at White River Junction, Vermont. The Two Cent Bridge, built by the Berlin Construction Company of Berlin, CT, a company established by former employees of the Berlin Iron Bridge Company after it was absorbed into the American Bridge conglomerate, was one of Edwin Graves' final bridges and it must have seemed to him a homecoming of sorts. Sadly, just three years after the Two Cent Bridge was built, he suffered a mental breakdown during the construction of the massive Bulkeley Bridge, a nine span stone arch bridge in Hartford, CT, which suffered four fatalities during its construction from 1903 to 1908, and he apparently never designed another bridge. Getting back to Waterville, the first foot bridge built here for the Ticonic Foot Bridge Company was built in 1901 and survived less than a year before being swept away by high water on December 15, 1901. Just over a year later, the current bridge was built. Intended as a toll bridge, it charged those passing over one cent in the beginning, with the toll house on the Waterville end of the bridge. The toll quickly rose to two cents, no doubt necessitated by the extra expenses incurred by the Ticonic Foot Bridge Company in building two bridges in less than three years, and it is this toll that gave the bridge its name.

The bridge remained a privately owned toll bridge until 1960, when it was given to the city of Waterville and the toll was subsequently abolished. Over the years the maintenance of the bridge has ebbed and flowed and at times it was closed due to its poor condition. However, waterfront redevelopment and reuse activities have given the bridge a new life over the past twenty years or so and now it is both a local landmark and a popular site for tourists. The cities of Waterville and Winslow and the civic institutions, recognizing the importance of the Two Cent Bridge, have provided funds over the years to keep the bridge in shape. The city's dual prominence as a center of industry and education in Maine is reflected in this bridge, which originally started as a factory bridge, but now is a recreational destination for locals and students alike. Interestingly, a concert along the banks of the Kennebec River in 1990 nearly brought the bridge down when hundreds of people gathered on it all at once. Severely damaged, the bridge had to be closed and took several years to repair. The distinctive toll house, which was once removed, has now been replaced as well, further adding to the bridge's historic nature. The Two Cent Bridge is currently in excellent condition and should continue to serve for many more years. The bridge is also unique in that it may be the only bridge in northern New England that has had a local beer named in its honor. The Mainely Brews Restaurant and Brewhouse, located just a short distance away in downtown Waterville, is a small craft brewery whose signature beer is named the Two Cent Bridge Ale, a good beer that honors the old bridge quite well!

16. Harrison — Ryefield Bridge; DATE BUILT: 1912; TRUSS TYPE: Warren double-intersection; FEATURE CROSSED: Crooked River; LENGTH: 98 feet; SPANS: 1; EASE OF ACCESS: Easy

This simple bridge is one of the most beautiful of the old metal truss bridges in all of New England. Carefully restored and painted a pleasing red color, the Ryefield Bridge is also

The Ryefield Bridge, Harrison.

significant as the only double-intersection Warren truss bridge remaining in Maine. While this bridge type was once popular in Maine from about 1900 to 1930, most were too lightly built to accommodate modern traffic and have subsequently been lost. This bridge, which was built using riveted connections and still has a wood planked deck, was built by the United Construction Company of Albany, New York, which was an arm of the American Bridge Company. Local residents recall that the bridge has been threatened by flooding down through the years, but has always survived intact. The Crooked River is usually a rather placid stream, but in springtime and during heavy periods of rain can become quite swollen and overflow its banks. Ryefield Bridge's historical status was cemented with its inclusion on the National Register of Historic Places in 1999. It is interesting to note that the town of Harrison had originally voted to build a steel bridge at the site of the Ryefield Bridge in September 1901, but within weeks rescinded the vote and instead voted for a new wooden bridge. Finally, in March 1912 the town voted to raise $1,000 to build a steel bridge and did so later that year. The town paid $1,565 for the bridge itself to United Construction, but also incurred an additional $47 in charges to haul the bridge components from the rail depot at nearby Norway.

---

17. Rumford — Morse Bridge; DATE BUILT: 1935; TRUSS TYPE: Three-hinged Steel Arch; FEATURE CROSSED: Androscoggin River; LENGTH: 285 feet; SPANS: 2; EASE OF ACCESS: Easy

The Morse Bridge is named for a once prominent business family in Rumford and replaced an earlier, more lightly built structure at or near the same site, not far from Rumford

The Morse Bridge, Rumford.

Falls. It is significant as the only steel arch bridge remaining in the state and was fabricated by the Pittsburgh-Des Moines Steel Company of Pittsburgh. Just a year after it was built, the Morse Bridge was damaged by the Flood of 1936, losing one of its abutments. Following this, the bridge was temporarily moved sideways on a track while a new abutment, steel-stringer approach span, and a cutwater pier were built in 1937–1938. Since that time, the bridge has well survived many other floods with few problems. The 230-foot long arch truss of this bridge is of the three-hinged type with box girder steel ribs from which are suspended steel H-section hangars that support its stringers and concrete deck. The portal bracing was altered and raised on the bridge in 1987 so that it could accommodate larger trucks. The area in which this bridge is located is one of great historical importance. It was at the pool below the Upper Falls, just downstream of the bridge on the Androscoggin River, that Native Americans of the St. Francis tribe gathered to hunt and fish for salmon along the riverbanks. Later, as might be expected, the falls area was the site of great industrial activity, which continues in Rumford today. Adding to the pleasant setting of the Morse Bridge is Memorial Park directly at the eastern end, while just a short walk away from the western approach is Morency Park, which is notable for its interesting Native American silhouette displays, some of which make for interesting viewing from the bridge.

---

18. New Sharon — New Sharon Bridge; DATE BUILT: 1916; TRUSS TYPE: Pennsylvania (pin-connected); FEATURE CROSSED: Sandy River; LENGTH: 268 feet; SPANS: 1; EASE OF ACCESS: Easy

This bridge was once the pride of the village of New Sharon, formerly carrying its main street over the Sandy River. The bridge was subsequently abandoned when a new concrete

This portal view of the New Sharon Bridge shows its steel grid deck and intricate overhead bracing system.

bridge was built to carry U.S. Route 2 over the river. The bridge early on was referred to as the Red Bridge, no doubt because of its bright lead paint color that was once common on early iron and steel bridges but fell out of favor by the 1920s. Now, the bridge is merely a reddish color from the rust that continues to corrode the bridge. Despite its threatened condition, the New Sharon Bridge is significant because of its pin-connected trusses. Though pin-connected bridges were once common, they fell out of favor beginning in the 1890s as riveted connections, which provided for a more rigid structure, took their place. It is unknown why the Groton Bridge Company used pin connections at this late date, though perhaps cost factors played a part in the decision. The Pennsylvania truss form employed by this bridge was also once very common, but is now an endangered species in Maine. With two other bridges of this type in the state slated for replacement, the New Sharon Bridge will soon become just one of three Pennsylvania truss highway bridges remaining in Maine, and the oldest. However, whether this bridge will survive is uncertain. Many in town would like to see it restored, but others view it as an eyesore. Older residents are fond of the old bridge and seem to have a greater appreciation for the soaring steel trusses that appear almost like a cathedral. Thus far, letters have been written by the town to the state DOT, as well as Maine Senator Olympia Snowe, but all their efforts have been in vain and local historians fear that it is only a matter of time before the bridge collapses into the river below due to neglect.

---

19. New Portland—The Wire Bridge; DATE BUILT: ca. 1866; TRUSS TYPE: Suspension; FEATURE CROSSED: Carrabasset River; LENGTH: 198 feet; SPANS: 1; EASE OF ACCESS: Easy

The village of New Portland in rural Maine, like most of northern New England, has little in common with such famed cities as New York and San Francisco. However, New Port-

The Wire Bridge, New Portland.

land *does* have one thing in common with them, a notable suspension bridge. True, the Wire Bridge is not as majestic as the Brooklyn Bridge or the Golden Gate Bridge, but what it lacks in size it more than makes up for it with its wilderness setting, its demonstration of Yankee innovation, and, most importantly, the fact that it is the second oldest highway suspension bridge in the entire country. The Wire Bridge is predated by only two other extant suspension bridges in America, only one of which, the Wheeling Suspension Bridge (built by Charles Ellet, Jr.) across the Ohio River, is a highway bridge. The other is the Delaware Aqueduct (built by John Roebling) across the Delaware River. While the building of these earlier spans, both constructed in 1849, and the careers of their famed builders are well documented, this is not the case in New Portland. It is now generally accepted by Maine bridge historians that the Wire Bridge was built ca. 1866–1868 by David Elder, who was paid $3,625 by the town (with no specifics) as bridge agent, and Capt. Charles Clark.

However, early records regarding this bridge are largely lacking and one local historian, Roland Foss, recorded in his 1950 history of New Portland a quite different scenario. He believed that the bridge had its beginnings in 1838, when the idea of building a bridge that would allow settlers on the north side of the Carrabassett River (also called Seven Mile Brook in this area) in the Parsons District to get to New Portland Village was discussed at the annual town meeting. Because of the high water brought on yearly by spring freshets, it was questioned whether a wooden bridge would hold up. However, one Col. F.B. Morse, an army veteran with engineering experience, suggested the idea of a suspension bridge, which he subsequently designed in 1840 and ordered the wire cables for the span from Sheffield, England. The cables were delivered across the Atlantic to Bath, Maine, shipped upriver via schooner to Hallowell, and were subsequently carried to New Portland in June 1841 by sixteen pairs of oxen under

A windshield view of the still heavily used Wire Bridge. Note the series of wire cable suspenders hanging from the main suspension cables which support the narrow wooden deck.

the direction of Moses Mitchell and Samuel Parker. The bridge abutments were built that summer by Ezra Wilson and William Witham under Morse's direction and the cables were strung in August, but work was later delayed on the bridge during the ensuing winter. However, the bridge, if this story is to be believed, was viewed with some skepticism and was even called by some Morse's "Fool Bridge." Finally, in June 1842 the bridge was opened, having cost $2,200. While this elaborate story is interesting, the problem is that there is no documentary evidence to back it up. The reason for building the bridge certainly makes sense, but there is no town record of bridge expenditures for the Wire Bridge this early on. Because of this lack of evidence for an earlier date, Maine DOT historians accept the 1864–1866 timeframe, but others have their doubts, citing the vagueness of the 1860s town records. A covered bridge was built in 1863 at East Village in New Portland, and is maybe this the structure that Elder and Clark were paid for? Some also believe that Roland Foss did not just make up this elaborate story and that the oral traditions on which it is based may have some credibility.

No matter which date you believe, the Wire Bridge is an unusual and significant bridge, and is one of at least four such bridges of this type built in this area of Maine. The Wire Bridge was probably inspired at least in part by the earliest known suspension bridge built in the state, built in 1852–1853 at nearby Kingfield by Daniel Beedy, but seems to have been modeled directly after the suspension bridge built at nearby Strong across the Sandy River in 1856. Interestingly, the cables for the Wire Bridge were not prefabricated elsewhere, but were actually spun in place, yet another strong indicator that Roland Foss' story about the building of this bridge, or at least that part about the cables being imported from England, is not wholly true. While the wooden deck of the bridge and the shingles coverings on its towers have been replaced over the years and the original iron rod suspenders which support the deck were replaced with wire cable suspenders, the bridge has been well maintained and undergone periodic repairs since the early 1960s and the timber framing of its towers, anchorage hardware, and the cables themselves remain original to the bridge. That the cables of the bridge have survived all these years is even more significant when we recall that the cables of the modern Waldo-Hancock Bridge on the Maine coast have deteriorated to such a degree that the bridge is now scheduled for demolition. Truly, the Wire Bridge is the epitome of a structure built to last. In recognition of its historic nature, the Wire Bridge was listed on the National Register in 1969 and was designated a landmark structure by the Maine section of the American Society of Civil Engineers (ASCE) in 1990.

---

20. Madawaska — International Bridge; DATE BUILT: 1921; TRUSS TYPE: Pennsylvania; FEATURE CROSSED: St. John River; LENGTH: 949 feet; SPANS: 4; EASE OF ACCESS: Moderate

An early view of the International Bridge at Madawaska.

The International Bridge, sometimes referred to as the Edmundston-Madawaska Bridge in deference to the Canadian city which it serves, is the most northerly of the metal truss bridges in New England and one of the most heavily traveled. It is also the oldest remaining historic truss bridge connecting Maine with Canada. The bridge was one of a number of bridges built as a joint effort between the Dominion of Canada and the bridge division of the Maine State Highway Commission beginning in 1910 and is unique in that its construction, because it involved our national border, had to be approved by the U.S. Congress. The town of Madawaska, on the Maine side of the border, is small, less than 5,000 inhabitants, most of whom are French Canadian in origin, while Edmundston, New Brunswick, lies on the Canadian side and is much larger, with about 16,000 people. The International Bridge, which carries U.S. Route 1 over the St. John River, was instrumental, and still is today, in not only facilitating travel in the area, but also in regards to business development. On both sides of the river are the pulp paper mills and related installations that are the economic lifeblood of the region. This bridge was the first major project for Maine bridge chief Llewellyn Edwards, who designed the structure, and one of the most important during his tenure at the bridge division from 1921–1928. The heavily built Pennsylvania truss design employed by the bridge, which replaced a cable ferry, was certainly suitable for a crossing such as this that would see heavy traffic and has served well over the years with few changes. To accommodate larger truck traffic, the portal bracing has been altered, while its original timber and asphalt deck was replaced with a steel grid deck in 1961.

The bridge carries over 750,000 vehicles a year between Canada and Maine and soon will be the only historic metal truss bridge between them. The 1927-built bridge at Fort Kent, Maine, also a Pennsylvania truss structure, has been tentatively scheduled for replacement beginning in 2012. Like many of the other metal truss bridges built between Maine and Canada, this bridge was fabricated by a Canadian company, in this case the Canadian Bridge Company. While Llewellyn Edwards designed the Madawaska Bridge, the engineers at Canadian Bridge were responsible for turning these designs into reality. One of their engineers, an African American, whose work on the bridge is undocumented, but nonetheless an intriguing possibility, was Cornelius Langston Henderson. Born in 1888 the son of the Reverend James Henderson of Detroit, Michigan, Cornelius would later move with his family to Atlanta, Georgia, where his father became president of Morris Brown College. As a young man, Cornelius attended Payne College, graduating in 1906. Five years later, he graduated from the University of Michigan with a degree in civil engineering, only the second African American to graduate with an engineering degree. Unable to find a job in his field due to the racial attitudes then prevalent in the United States, Henderson accepted a position as a draftsman with the Canadian Bridge Company in 1911, located just across the river from Detroit in the Walkerville section of Windsor, Ontario. Cornelius Henderson worked as a draftsman for the first four years of his career, but soon moved up in the company. While he is most noted for working on the famed Ambassador Bridge (1928) connecting Detroit and Windsor, and the Detroit-Windsor Tunnel (1930) project, Henderson undoubtedly gained much valuable experience on earlier bridge projects. One of these was likely the International Bridge project at Madawaska; it seems probable that Henderson, nearly a ten year veteran with Canadian Bridge by this time, played a part in working with Edwards' design specifications at Canadian Bridge's Walkerville plant. Involvement with such projects as the International Bridge at Madawaska would certainly have been a natural progression in Henderson's development as a bridge engineer and designer prior to the Ambassador Bridge project.

13

Interstate Bridges

1. Portsmouth, New Hampshire, and Kittery, Maine — Memorial Bridge*; DATE BUILT: 1921; TRUSS TYPE: Vertical lift, Warren truss; FEATURE CROSSED: Piscataqua River; LENGTH: 1,201 feet; SPANS: 3; EASE OF ACCESS: Easy

The Memorial Bridge, Portsmouth, New Hampshire, Kittery, Maine.

Of all the bridges in this survey, the Memorial Bridge is considered to be the most historic, as well as the most endangered. Indeed, during much of the 1980s and 1990s, the potential fate of this aging bridge was heavily debated, and continued to build in the first decade of the 21st century, culminating in its designation in 2009 as one of America's eleven most endangered historic places by the National Trust for Historic Preservation. However, speculation about rehabilitating the old span ended in October 2010 when the formal decision was made by New Hampshire and Maine to demolish the structure and build a new bridge in its place.

The bridge, which consists of three Warren truss spans each measuring 297 feet long,

This view shows the elaborate memorial plaque and seals on the New Hampshire side of the bridge.

was begun in 1921 and is named in honor of the World War I servicemen from New Hampshire and Maine that fought in the conflict. In keeping with its purpose as a memorial, the portal of the bridge at either end has a bronze memorial plaque with wording to this effect flanked with the state seal and the Great Seal of the United States, while above is a majestic eagle, and below the plaque is a union shield. This grouping of symbols was designed by the Gorham Company of Providence, Rhode Island, nationally famed for their work in silver and pewter.

The Memorial Bridge is the first bridge at this spot between Portsmouth and Kittery, Maine, replacing an old wooden toll bridge located about a mile upstream. The need for the bridge was agreed upon not only by the two states that it connects, each contributing $500,000 of the cost, but also the U.S. Navy, which contributed the final $500,000 to finance the bridge. Not only does the bridge to this day serve as a vital link to the civilians working at the Portsmouth Naval Shipyard, located just east of the bridge on the Maine side, but it also directly served the interests of the navy, initially enabling supplies to be shipped to the naval yard much more directly and at a lesser cost, while also allowing for quick access by the Portsmouth Fire Department in case of emergency. Today, the bridge is vital to the interest of tourism in both states, as well a providing a local link between the downtown areas of both cities used by motorists, bicyclists, and pedestrians alike. Memorial Bridge is also notable for its bridge type and the nationally known engineer, J.A.L. Waddell, who designed the structure. Not only was the lift span of this bridge (297 feet) the longest in the country at the time, but it was also the largest such bridge on the East Coast. Waddell had previously built similar spans only in the Midwest, starting with Chicago's South Halstead Street Bridge, and such notable western bridges as the Hawthorne Avenue Bridge in Portland, Oregon. While later and larger vertical lift bridges would soon be built in the coming years, Memorial Bridge, the 39th bridge built by Waddell, is unique today in that it is one of the oldest vertical lift bridges in the country that is still operational and largely unaltered.

The main change to the bridge came in 1977, when the original machinery, such as motors, brakes, gear boxes, and winding drums, was fully replaced. Once exposed to view, this machinery is now enclosed and its moving parts, so to speak, are largely hidden. The bridge also once had a wooden deck that has now been replaced with an open steel grid deck, and the noise made when cars and trucks go over it have led some to call Memorial Bridge a "singing bridge." The use of the Warren truss in Memorial Bridge was unique for Waddell at the time, as he generally used the Pratt truss for his lift bridges. Several other unique and little known aspects of Memorial Bridge include the fact that its vertical lift clearance of 150 feet at high tide is one of the highest of any lift bridge in the country. This requirement was made by the U.S. Secretary of War at the time of its building, Josephus Daniels, to accommodate the high volume of merchant sail traffic, primarily coal schooners (including some giant six-masters), then active in plying the waters of the Piscataqua River. Today, the largest ships that the bridge must accommodate include modern bulk cargo and container ships, cruise ships, and even, on occasion, a gathering of historic tall ships. It is also a little known fact that Memorial Bridge came close to being a combination railway and highway bridge, and was in fact constructed to accommodate both types of traffic at the same time. This might have happened, but the Portsmouth, Dover, and York Street Railway (an electric streetcar system) went out of business before the bridge was finished in 1923, and since its inception Memorial Bridge has always been best known for carrying U.S. Route 1 over the Piscataqua River.

Of course, Memorial Bridge has always been a feature on the Portsmouth skyline since its building. Its 200 foot high towers, which support the counterweights for the lift span and through which run its sixteen wire cables, have blended in quite nicely with such other historic structures as North and St. Johns churches. The bridge was probably even more imposing when it was first built, as it was painted black, the usual color for Waddell's bridges. This paint scheme lasted until after World War II, when the bridge gained the green paint job that was standard with both the state's bridge divisions. The operator's house for the Memorial Bridge, as was typical of Waddell bridges, is located on the center span in the middle, about nineteen feet above the roadway. Not to be overlooked are the pier supports for Memorial Bridge and the approach spans on each side of the three main spans. The Piscataqua River is one of the fastest flowing in the world and constructing the piers of the bridge was one of the main challenges, where the water was about 60 feet deep and timber caissons were utilized. The piers, partly sheathed in granite, measure about 75×25 feet at the bottom of the river, where they were set in a rock ledge once debris in the area had been removed, and taper to about twelve feet wide at the top, where they were capped with concrete. As to the approach spans, as is typical with such large structures, there are many; ten steel deck girder spans on the Maine side and five reinforced concrete slab spans on the New Hampshire side.

Memorial Bridge was heralded upon its completion in 1923 as one of the nation's most modern bridges and has served the cities of Portsmouth and Kittery well for over eight decades now. However, by 2005 the structure had deteriorated to such a degree that decisions had to be made about the bridge's fate. It was originally thought in 2008 that the bridge could be saved with funds from both Maine and New Hampshire, estimated to be about $90 million, but these plans fell through. While New Hampshire pledged her share of the repair costs, Maine dragged its feet and, despite conducting some feasibility studies on their own, never really committed to the cause of saving the bridge, much to the chagrin of Kittery residents. A further blow to efforts to save the bridge came when federal funds applied for under President Obama's Economic Stimulus program were rejected. The hope for preservation of the bridge ended in October 2010 when U.S. Transportation Secretary Raymond LaHood delivered a check for $20 million to the New Hampshire Department of Transportation to fund the demolition of Memorial Bridge and its replacement with a new bridge. As of this writing, the new bridge is in its early planning stages, but construction is slated to begin in June 2012 and will last at least 18 months. Interestingly, the first design of the new bridge was largely patterned after that of Memorial Bridge, but many in the public rejected it, desiring a more modern looking span. While Memorial Bridge has been closed at various times due to maintenance issues, the final blow may have come to the old bridge when it was closed on December 9, 2010, after undergoing a periodic inspection. While the bridge may never carry highway traffic again (though temporary repairs are being explored), it still is open for pedestrian traffic, and

Well lit at night for both marine navigation and as a warning to local air traffic, the Memorial Bridge is a prominent feature on the nighttime skyline in Portsmouth and Kittery.

the vertical lift span remains operational. While many in the community may debate the decided fate of Memorial Bridge, no one can deny its historical importance and its demolition is sure to be both a spectacular and sad event.

---

2. Portsmouth, New Hampshire, and Kittery, Maine — Sarah Mildred Long Bridge; Date Built: 1940; Truss Type: Vertical lift, Warren; Feature Crossed: Piscataqua River; Length: 2,804 feet; Spans: 26; Ease of Access: Difficult

The youngest of the historic vertical lift bridges remaining in northern New England, the Sarah Mildred Long Bridge sometimes seems to be a forgotten span, overshadowed by the Memorial Bridge just downstream. This bridge, which gained its current name in 1988 after a long-time secretary who worked for the Interstate Bridge Commission, cost about $3,000,000 when it was built in 1940 and served two purposes that it still fulfills today. The first of these was to mitigate the traffic congestion in downtown Portsmouth and Kittery caused by backups on U.S. Route 1 as it crosses Memorial Bridge by bypassing the area further upstream, thus creating the U.S. Route 1 Bypass section of the national highway system. The other purpose of this double-decked bridge was to carry the tracks of the Boston and Maine Railroad over the Piscataqua River. By doing so, it replaced an antiquated wooden railroad bridge just a short distance away, while at the same time eliminating another obstruction to navigation by combining two bridges into one. This new railroad bridge couldn't quite come soon enough

This view shows the Sarah Mildred Long Bridge in the foreground, less than twenty feet above the Piscataqua River on its lower deck, with the high-rise I-95 Bridge in the background. Contrary to the thoughts of many, one bridge between New Hampshire and Maine in this heavily traveled area is not enough, and the Sarah Mildred Long Bridge will remain in service.

for the Boston and Maine, as one of their trains, locomotive #3666 built by the Baldwin Locomotive Works, plunged through the old bridge on September 11, 1939, killing two trainmen. The wreck lies underwater close to the Sarah Mildred Long Bridge to this day, having never been salvaged. Today, the rail line carried by the bridge is one of great significance to the U.S. Navy, as it serves the nearby Portsmouth Naval Shipyard and provides for the shipment of waste and other materials from overhauled and decommissioned nuclear submarines worked on at the yard.

The Sarah Mildred Long Bridge is also significant for its nationally known designer, as well as the fact that it represented some of the latest advancements made in the field of vertical lift bridges. The designing firm was that of Harrington and Cortelyou of Kansas City. John Harrington was a student, and later partner with the nation's premier moveable bridge engineer, J.A.L. Waddell, and the two men together were responsible for many of the advances in moveable bridge technology in the first decades of the 20th century. The Sarah Mildred Long Bridge was advanced for its time, one of the last vertical lift bridges built before World War II, and one of only a handful built in the 1940s and 1950s. The most notable improvement incorporated into the design of this bridge was the positioning of its drive and synchronous 100 hp motors at the top of the towers themselves, rather than in an operator's house on the lift span, eliminating the use of cables to transfer power from the house to the towers. This made for a much more efficient lifting system and one generally requiring less maintenance. It was first used on the Buzzards Bay Bridge over the Cape Cod Canal in Massachusetts in 1935. The superstructure of this bridge was fabricated by the Phoenix Bridge Company of Phoenixville, Pennsylvania, its main spans consisting of a 245-foot long Warren truss lift span, flanked on each side by 227-foot long Warren deck truss spans. The remaining approach spans for the bridge consist of fifteen deck girder spans on the Portsmouth (south) side skewed to allow for a horizontal curve, while on the Kittery side are six deck girder spans ranging between 70 and 90 feet in length. As to its artistic merits, the Sarah Mildred Long Bridge was built in the Art Moderne style popular in the 1930s, a stripped down version of the Art Deco style from which it sprang, emphasizing a more modern and streamlined look first developed by industrial designers. This same influence is also present in the bridge's concrete abutments.

Like many large bridge projects of the 1930s, this one too relied on federal aid, in this case from the Public Works Administration, one of President Roosevelt's New Deal programs. Like its downstream neighbor, this bridge is also in need of significant work, as its superstructure is rusting and its deck deteriorated. For some time prior to 2010 the fate of this bridge and its possible demise altogether was tied up with the fate of Memorial Bridge. After all, was there a need for two vertical lift bridges so close to one another, especially now that the high rise bridge carrying I-95 over the Piscataqua River already carried significant traffic once carried by the Sarah Mildred Long Bridge? The issue was debated for some years, with the slight possibility that the Sarah Long Bridge, the less historic of the two vertical lift bridges, might be sacrificed and replaced with a new hybrid bridge, similar to the 1912-built double-lift Steel Bridge in Portland, Oregon, whose higher road deck would result in 70 percent fewer lifts. However, in December 2010 it was formally decided that the bridge would be retained and rehabilitated, after the demolition of Memorial Bridge and the building of a new bridge built to take its place. With this joint decision by the states of New Hampshire and Maine, the Sarah Mildred Long Bridge will soon, by 2015, become the grand dame of Piscataqua River bridges.

---

3. Stewartstown, New Hampshire, and Canaan, Vermont — Beecher's Falls Bridge; DATE BUILT: 1930; TRUSS TYPE: Two-hinged steel deck arch; FEATURE CROSSED: Connecticut River; LENGTH: 232 feet; SPANS: 1; EASE OF ACCESS: Easy

The Beecher's Falls Bridge, Stewartstown, New Hampshire, and Canaan, Vermont.

A bridge has stood at this spot since 1885, when the citizens of Beecher's Falls, Vermont, raised funds to the tune of $1,500 to build a covered bridge. This bridge stood, in a very dilapidated state in its later years, until the current bridge was built in 1930. The choice of an arch deck bridge for the new structure was surely influenced by the logging and lumber industry practiced on both sides of the river, allowing log-laden trucks to easily cross the river without the possible limitations of a through-truss bridge. Today, the logging and related industries are much reduced and tourism has taken its place. The two-hinged arch bridge was designed by Harold Langley of the New Hampshire Highway Department and was fabricated by the American Bridge Company. It was erected by the Kittredge Bridge Company of Concord, New Hampshire, a bridge construction firm that was active in the state during the 1930s. The bridge was subsequently recognized as the Most Beautiful Bridge among class C bridges in 1931 by the American Institute of Steel Construction. In early 2010 the bridge was closed for four days to undergo a thorough inspection.

4. Stratford, New Hampshire, and Maidstone, Vermont — Janice Peaslee Bridge; DATE BUILT: 1893; TRUSS TYPE: Pratt through (pin connected); FEATURE CROSSED: Connecticut River; LENGTH: 152 feet; SPANS: 1; EASE OF ACCESS: Easy

The Janice Peaslee Bridge, Stratford, New Hampshire, and Maidstone, Vermont (courtesy of Craig Hanchey).

Built in 1893, this bridge is the oldest pin-connected truss bridge still in use today in New Hampshire, and the oldest bridge across the Connecticut River between New Hampshire and Vermont. It is also interesting in that it is fashioned, at least in part, of wrought iron instead of steel, which was coming into rapid use by this time, making it thus even more of a throw-back to the early days of metal bridge building. Formerly called the Stratford Hollow

Bridge, this bridge's name was officially changed to its current name with the passage of New Hampshire Senate Bill 254 to honor the long-time Vermont state legislator, a resident of Guildhall, who was instrumental in helping to get the bridge restored. Its original construction was also the work of a politician, initiated by Stratford's representative to the N.H. legislature, Fred Day, in 1893; he also pushed for, and got, approval for another bridge in North Stratford. Whoever has said that ear-mark, or pork-barrel spending by the government is all bad probably never lived in a town that really needed a bridge! Day was also successful in getting a resolution passed that the state would pay one-third of the cost of both bridges up to the amount of $4,000. The Berlin Iron Bridge Company of East Berlin, CT, won the contract to build both bridges for $16,065, beginning construction on both in November 1893 and finishing by May 1894. The town of Maidstone's share in the project was but $500. Interestingly, the larger and more costly lenticular truss bridge across the Connecticut from North Stratford to Bloomfield, Vermont, is now long gone, having been replaced in 1947. Its contemporary, a smaller and much more practical span than its northern neighbor, employing a solid Pratt truss design, has now served the region well for nearly 120 years now, including local residents and tourists alike.

However, the recognition of the importance of this bridge to the people of Stratford and Maidstone, and its eventual restoration, took many years to achieve. In the 1950s the state of New Hampshire gave up the maintenance of the bridge and turned it over to the town. The small community was, however, unable to fund the bridge's upkeep, and, as resident Bill Paradis thought, did not spend one cent to do so. Finally, in the early 1970s the town voted to abandon the bridge. However, some residents were not willing to let it go and were concerned about the effect the lack of a crossing might have on the two towns. Three residents, Bill Paradis, Jim Emerson, and Sally Ryan, formed the Stratford Hollow-Maidstone Bridge Association and bought the bridge from the town for $1. With money hard to come by to fix their now privately owned bridge, the owners decided in September 1974 to set up temporary toll gates at each end of the gate to charge a dollar toll from 7:30 a.m. to 5 p.m. on Labor Day, and subsequently on several other holidays. On that day alone $400 was collected, a direct reflection on how much the bridge was really used. While most people gladly paid the toll, Bill Paradis would later recall, "Of course there were some people who were so doggoned cheap that they backed up and went the other way" (Harrigan, *New Hampshire Sunday News*, June 15, 1975). With the money raised, the owners, along with friends, family, and local businesses, worked to put new steel H-beams in the floor, replace the wooden deck, and scrape and paint portions of the bridge. Though more work was needed, Paradis and the members of the bridge association kept going. While periodic tolls were collected, the thought of everyday tolls was never seriously considered due to the hardship it would impose on local farmers and workers that used the bridge daily. Ownership of the bridge was difficult at times, but Paradis, Emerson, and Ryan refused to quit. Bill Paradis, indeed, was adamant, stating that "a lot of people have got discouraged and have just quit trying to keep the bridge. So many people just sit back and do nothing.... Hell, that bridge is absolutely vital to this area.... For farmers and other residents of Maidstone, the nearest fire department is here in Stratford Hollow, and so is the nearest ambulance.... If that bridge saves just one life, in my book its paid for itself" (*ibid.*). Despite the Herculean efforts of these few private citizens, Bill Paradis in particular, they could not keep up with the bridge's increasing maintenance needs and it had to be closed due to its unsafe condition.

Finally, in 2006 the bridge was restored with new concrete abutments, new floor supports and decking, and a new paint job, as well as an extension on the Vermont side to allow better water flow. The area around the bridge was further enhanced with the addition of a public access area for fishing and boating, the granite blocks from the original abutments now lining the path down to the Connecticut River. In short, the rehabilitation of the bridge was the result of outstanding cooperation of agencies from New Hampshire and Vermont, as well as citizens and politicians on both sides of the river, and finally, regional groups like the Connecticut River Joint Commission and the Northwoods Stewardship Center. Today, those original folks that tried to save the bridge back in the 1970s, visionaries like Bill Paradis, have all passed on. However, it was their efforts, these patron saints of the bridge, if you will, that bought this old span some time before those outside Stratford and Maidstone finally got their act together and came to recognize the bridge for the important and historic structure that it is.

---

5. Dalton, New Hampshire, and Lunenberg, Vermont — Whitcomb Bridge; DATE BUILT: 1928; TRUSS TYPE: Pratt deck; FEATURE CROSSED: Connecticut River; LENGTH: 547 feet; SPANS: 3; EASE OF ACCESS: Easy

The Whitcomb Bridge, Dalton, New Hampshire, and Lunenberg, Vermont (courtesy of Craig Hanchey).

Now closed to traffic, this bridge on Gilman Road was either named in honor of the family by that name in Dalton, including one of the selectmen in office when the bridge was erected whose name is inscribed on its plaque, or one of the area's heroes during the American Revolution, Col. Benjamin Whitcomb. A Massachusetts native, Whitcomb later moved to the New Hampshire Grants, living in nearby Guildhall before the war and afterwards settling down in Lisbon, New Hampshire. Whitcomb's Rangers were instrumental in fighting the British and their Indian allies in northern New Hampshire and Vermont during the war, achieving such great success that a bounty was placed on Whitcomb for his capture. Local residents I spoke to during a visit to the area offered both versions for the naming of the bridge, with the thought that the latter day selectman was a descendant of Benjamin Whitcomb. A wooden bridge once stood in this area, but had gone out some years prior to 1900 and details about it are lacking. The three span Pratt deck truss Whitcomb Bridge that now stands here was built by the Berlin Construction Company and is important as the longest historic span of its type in the state of New Hampshire. The bridge was bypassed in 1998 when a new bridge was built just downstream and now serves as a pedestrian crossing. However, the bridge has had but little maintenance over the years since it was closed to vehicles and it is now in need of a new paint job and repairs to its deck.

---

6. Monroe, New Hampshire, and Barnet, Vermont — McIndoe Falls Bridge; DATE BUILT: 1930; TRUSS TYPE: Parker through; FEATURE CROSSED: Connecticut River; LENGTH: 305 feet; SPANS: 1; EASE OF ACCESS: Easy

The McIndoes Falls Bridge, Monroe, New Hampshire, and Barnet, Vermont.

This site on the Connecticut River, like so many others, has been bridged constant times over the years. The first bridge, a wooden span of an unknown type, was built in 1803 and lasted for thirty years before being brought down by flood waters. Its successor was a two-span covered bridge, known as Lyman Bridge, which operated initially as a toll bridge and was in use until the new bridge was completed directly adjacent downstream, just before the falls. The current bridge is an attractive Parker truss span fabricated by the American Bridge Company and designed by the state highway department. In March 1930 the town voted to raise $27,500 to build the bridge, borrowing $12,000 of that amount and raising $15,500 by tax levies, while the Connecticut River Development Corporation agreed to reimburse the town any costs exceeding $35,000. The building of this bridge was significant not only locally, but on a regional level as well. It was part of the massive Fifteen Miles Falls Development Project on the Connecticut River by the New England Power Company begun in 1928 with the construction of the huge Comerford Dam upstream in North Monroe, and continued with the building of the small, but important McIndoes Falls Station in 1931, followed by yet another dam further downriver in Littleton in 1957. The dam at McIndoes Falls is one of eight power generating stations on the Connecticut River, and measuring 25 feet high is the smallest. However the dams of the Fifteen Mile Project, now owned by the TransCanada Corporation, are said to be the largest hydroelectric power generating complex in New England. The replacement of the old covered bridge at the site of McIndoes Falls was required because the piers of the two-span bridge served to restrict the flow of water in the Fifteen Miles area and likely would not have withstood the increased water flow. The current bridge was begun in April and completed by September 1930, with construction on the dam well underway by this time. The McIndoes Falls Bridge was fully rehabilitated in 2006.

---

7. Haverhill, New Hampshire, and Newbury, Vermont — Woodsville-Wells River Bridge; DATE BUILT: 1923; TRUSS TYPE: Steel Arch; FEATURE CROSSED: Connecticut River; LENGTH: 259 feet; SPANS: 1; EASE OF ACCESS: Easy

The Woodsville–Wells River Bridge, Havenhill, New Hampshire, and Newbury, Vermont.

Situated directly downstream from the combination highway and railroad bridge, this bridge is the oldest steel arch span on the Connecticut River and the most beautiful. It was designed and built by the famed engineer, J.R. Worcester, of Boston, who had previously built the first arch bridge, and the longest in America at the time, across the Connecticut River further downstream at North Walpole. The bridge components of this three-hinged arch were fabricated by Boston Bridge Works, noted for their large steel spans. The Woodsville-Wells

River Bridge was a flood-replacement bridge in a year not generally noted for its massive flooding, succeeding the 1917-built Ranger Bridge, a three-span Warren deck truss bridge which was designed by John Storrs and cost nearly $65,000. Just five years after it was built flood waters undermined the piers of the Ranger Bridge, destroying the entire structure, a rare failure for Storrs. The new bridge by Worcester, taking the cause for this failure into account, resorted to an arch structure at this location, thus avoiding the need for pier supports that could potentially be undermined during extreme flood conditions. This vital link between New Hampshire and Vermont has now performed admirably for nearly 90 years and, with its rehabilitation in 2001–2003, should be good for many more years of service.

---

8. Haverhill, New Hampshire, and Newbury, Vermont — The Old Woodsville–Wells River Bridge; DATE BUILT: 1903; TRUSS TYPE: Baltimore (pin-connected); FEATURE CROSSED: Connecticut River; LENGTH: 253 feet; SPANS: 1; EASE OF ACCESS: Easy

The old Woodsville–Wells River Bridge as it appears today. Note the pin connections at the center of each truss panel and the steel support beams for the lower highway deck that are mounted above the lower chord of the bridge.

To look at this abandoned bridge today, you might think you're looking at nothing but a rusting hulk, which, to tell the truth, is what it could become. However, in its heyday this unusual combination railroad and highway toll bridge was quite a sight to behold, and was very important to both the local and regional economy and transportation network. While the first bridge built at this site was built in 1805, this bridge replaced an earlier covered com-

Connecticut River Toll Bridge, Woodsville, N. H.

This postcard view not only shows the railroad traffic that ran on top of the old Woodsville–Wells River Bridge, but also the dangerous highway approach to its lower level.

bination highway and railroad bridge built in 1853 by the Boston, Concord, and Montreal Railroad. The Boston and Maine Railroad would later take over this route (even later partnering with the Canadian Pacific Railroad) and commissioned this new Baltimore truss steel bridge to be built in 1903. It is an interesting survivor because of its pin-connected construction, once common but now rare among surviving bridges. However, the use of pin-connections rather than rivets by this time seems an odd choice for a bridge that would carry both highway and rail traffic, and this at a time when most railroads were building riveted steel spans. Whatever the reason, the bridge did its job for years without any problems. Like its wooden predecessor, trains traveled on the top of the bridge, while highway traffic, which had to pay a toll, traveled beneath, a steel version of the popular wooden bridges of the same type, utilizing the Howe truss and often called "upside down" bridges, of a previous generation. It must have been quite a thrill, and very noisy, when traveling in a car or horse and buggy when the Boston and Maine's trains to Montreal, later called the *Alouette* or *Red Wing* express, thundered over the bridge!

Because of the bridge's approach alignment, requiring a sharp curve near the entrance for highway traffic, highway traffic was discontinued when the new Ranger Bridge was built just downstream in 1917. The Woodsville–Wells River Bridge would remain a railroad bridge for many years longer until all service ended by the late 1970s, but automobiles would not be carried by it again until 84 years later. By 2001, the railroad tracks to and from the bridge were largely gone and the bridge, though for all intents and purposes abandoned, was refurbished for temporary use by auto traffic on its upper deck while work on the adjacent highway bridge was being done. Once the arch bridge was repaired, this bridge has now gone back to its abandoned state, though it appears it could be used again for temporary highway traffic should the need ever arise, at least in the short-term future.

9. Piermont, New Hampshire, and Bradford, Vermont — Piermont Bridge; DATE BUILT: 1928; TRUSS TYPE: Pennsylvania; FEATURE CROSSED: Connecticut River; LENGTH: 355 feet; SPANS: 1; EASE OF ACCESS: Moderate

The Piermont Bridge, Piermont, New Hampshire, and Bradford, Vermont.

This impressive span is the longest Pennsylvania truss bridge in the state of New Hampshire, and was the state's largest flood-replacement project in the wake of the Flood of 1927. It replaced a two-span Town lattice covered bridge that was built in 1877 as a toll bridge by the Piermont Bridge Company and served as such until Piermont and Bradford freed the bridge in 1901 by paying $6,000. The new bridge, which carries New Hampshire State Route 25 over the Connecticut, was fabricated by the Boston Bridge Works and designed by the New Hampshire Highway Department's bridge division. Interestingly, it retains the covered bridge's concrete eastern abutment, which was designed by John Storrs in 1908, and utilized the surviving central pier of the wrecked covered bridge during its construction which was afterwards removed. Located in a bucolic setting, where cows can often be seen grazing in a pasture on a farm adjacent to the site, the narrow bridge carries a fair amount of traffic and is well maintained. It is interesting to note that even way back in 1938 one Connecticut River traveler noted about this bridge that "a fine herd of cattle happened to be in line with our picture and added much to it" (Whittlesley, pg. 97). Some things never change!

10. Orford, New Hampshire, and Fairlee, Vermont — Samuel Morey Memorial Bridge; DATE BUILT: 1937; TRUSS TYPE: Steel Arch; FEATURE CROSSED: Connecticut River; LENGTH: 434 feet; SPANS: 1; EASE OF ACCESS: Moderate

The Samuel Morey Memorial Bridge between Orford, New Hampshire, and Fairlee, Vermont, is here depicted in this early Richardson photograph postcard.

This bridge was yet another flood-replacement bridge, built by the state of New Hampshire in the wake of the Flood of 1936, although far from your ordinary bridge. The site was first bridged in 1800 with a wooden arch bridge, which only survived for nine years until flood waters brought it down. It was subsequently replaced with an open-decked wooden bridge supported on three piers that lasted until 1856. The third bridge was a two-span Town lattice covered bridge built by James Tasker and Bela Fletcher which served as a toll bridge until it was freed in 1896. It continued to serve until the floodwaters of March 1936 brought it down for good.

A box-girder, tied arch bridge design was chosen for the new bridge and designed by New Hampshire bridge division engineers John Wells and R.S. Burlingame under the guidance of assistant bridge chief, Harold Langley, a noted expert on steel arch bridges. The arch design was a good choice for the new bridge, which has since then well survived seasonal flood waters. Not only did its design pay homage to the original wooden arch bridge, but it also avoided the need for pier supports that constricted water flow and could be undermined in extreme conditions. Interestingly, the Samuel Morey Bridge is named after the man that was the first to operate a steam-powered boat on the Connecticut River at this location in 1793. Plans were made in late 1936 for the new bridge and the project was subsequently put up for bidding by early December. The contract was soon awarded to the Hagen and Thibodeau Construction Company of Wolfeboro, New Hampshire, at a cost of $209,631. By the first week in January 1937, work was begun on the bridge site. Meanwhile, the bridge superstructure, which cost $153,100, was fabricated at the Ambridge plant in Pennsylvania and in the summer of 1937 was shipped to Fairlee, Vermont, via the Boston and Maine Railroad. From the town depot the eight preassembled boxed-girder sections were transported to the bridge site via dollies pulled by a Mack truck — quite an event, the likes of which had never before witnessed in this small village! The sections were later hoisted into place and cantilevered out over the river, held in place by cables and timber falsework until the arch was fully in place. The Samuel Morey Bridge was completed on June 13, 1938, and dedicated in an elaborate ceremony some two weeks later. Later in the year the bridge was awarded second place among Class C bridges by the American Institute of Steel Construction in its annual Most Beautiful Bridge competition for 1937. The first place winner was the John Wells-designed arch bridge built further downstream at Chesterfield. Today, the Samuel Morey Bridge has been well cared and not long ago received a new paint job, the shade of the green decided on having been the subject of much historical debate, as has been previously discussed at length.

---

11. Charlestown, New Hampshire, and Springfield, Vermont — Cheshire Bridge; DATE BUILT: 1930; TRUSS TYPE: Pennsylvania; FEATURE CROSSED: Connecticut River; LENGTH: 489 feet; SPANS: 3; EASE OF ACCESS: Easy

The Cheshire Bridge is yet another current Connecticut River span that occupies a historic site, where rights to build a bridge were first granted in 1804. Its immediate predecessor was a three span Pratt steel truss bridge built by the Berlin Iron Bridge Company in 1906 at a cost of $65,000, replacing a Town lattice covered toll bridge. This first steel span was also a toll bridge that carried both highway traffic and the tracks of the Springfield Electric Railway Company over the river, utilizing the piers of the old covered bridge. The current Cheshire Bridge also utilizes the piers of the covered bridge and was fabricated by the steel giant McClintic-Marshall, costing $225,000. You wouldn't know it today, but the area of this bridge,

The Cheshire Bridge, Charlestown, New Hampshire, and Springfield, Vermont (courtesy of Craig Hanchey).

and the short rail line that led to nearby Springfield, Vermont, was once a center of industry and even a tourist destination; an amusement park was located near the Charlestown end of the bridge in its heyday. Today, the area around the bridge is a quiet and rural setting, though the crossing itself is an important link between the communities of Charlestown and Springfield and provides quick access for locals and tourists alike to three major roads, U.S. Route 5 and Interstate Route 91 in Vermont and State Route 12 in New Hampshire.

12. Chesterfield, New Hampshire, and Brattleboro, Vermont — Chief Justice Harlan Fiske Stone Bridge; DATE BUILT: 1937; TRUSS TYPE: Steel Arch; FEATURE CROSSED: Connecticut River; LENGTH: 440 feet; SPANS: 1; EASE OF ACCESS: Easy

This postcard view shows the old bridge at Chesterfield as it originally appeared.

Though now bypassed, this newly renamed bridge was once one of the most celebrated of all the Connecticut River bridges between New Hampshire and Vermont. The bridge was the largest of the Connecticut River flood replacement bridge projects undertaken by New Hampshire after the Flood of 1936, though it might be best termed a semi-flood replacement bridge. It replaced a Berlin Iron Bridge-built suspension bridge that was badly damaged, not in the flooding of 1936, but when a large truck had passed over the bridge in early November 1935, buckling the structure to such an extent that guards were placed at each end of the bridge and nothing heavier than a single automobile at one time was allowed to pass. Temporary repairs allowed the bridge to be continued in use for light traffic, but both Vermont and New Hampshire agreed that the bridge needed to be replaced. The matter with the old bridge was settled on March 19, 1936, when flood waters finished the job by destroying the already damaged bridge.

Previous plans for the replacement of the Chesterfield Bridge must have already been underway before the floods, for plans for the new bridge were soon drawn up and by early May, with federal aid targeted for the new bridge, a single-span steel arch bridge design was chosen. The bridge was engineered by John Wells and R.S. Burlingame under the direction of New Hampshire assistant bridge chief Harold Langley, all of whom were very busy in 1936–1937 on a number of flood replacement bridges. It was similar in nature to the Samuel Morey Bridge upriver at Orford. Indeed, the process of designing the Chesterfield Bridge was a timesaver later on, as some of its design drawings and calculations were reused on the planning of the Orford span. Construction on the two-hinged, half through arch, estimated to cost $125,000, was begun in November 1936 by the O.W. Miller Company of Springfield, Massachusetts.

The abutment on the Vermont side was first excavated and worked on through the winter using heated concrete (poured at 80 degrees and heated inside the abutment form by oil burners), while the New Hampshire side of the work also proceeded apace and was completed by March 1937. The bridge components were fabricated by Bethlehem Steel Corporation in Pennsylvania and shipped via rail to Brattleboro, where it was carried by heavy trucks across a temporary wooden bridge to the New Hampshire shore. Construction of the arch began in April 1937 with the construction of falseworks to provide support for the boxed arch rib sections, eight in all and each weighing 22 tons, as they were cantilevered out over the water. Traveling derricks were used to hoist the rib sections in place one by one. Once the arch was completed, the hangars and beams to support the deck were added, and the deck stringers were laid. The bridge construction proceeded with only a few complications, none of them readily foreseeable. There was a brief strike by nearly 20 steel workers, who walked off the job demanding a pay raise of 30 cents to $1.50 an hour. However, Bethlehem Steel refused this request and, because they weren't union workers, most of them decided to return to work, their demands unfulfilled. More dangerous was some unexpected flooding in May 1937 that destroyed the temporary wooden bridge and threatened the falsework. However, the bridge was nearing its final weeks of construction and would not be denied. The final arch ribs were put in place in June and on July 31, 1937, the bridge was open to traffic. Though there was no formal dedication service, the Chesterfield Bridge was still thronged with motorists, all hoping to be the first to cross the bridge!

The final cost of the bridge itself had reached $190,000 according to local news reports, but Langley would later write that the cost was $135,000 (Hool and Kinne, pg. 374c). Local news reports also recorded that roadwork on the approaches cost about $85,000, probably an accurate figure, mainly because the site of the bridge had been removed just upstream from the old suspension bridge. The bridge was a true success in every way, providing a solid and reliable bridge for many years to come and one that was aesthetically appealing, winning an award as the Most Beautiful Steel Bridge in Class C for 1937.

The Chief Justice Harlan Fiske Stone Bridge (left) and its replacement, Chesterfield, New Hampshire, and Brattleboro, Vermont.

The bridge was bypassed in 2003 with the completion of a new arch bridge directly upstream, almost side-by-side with the original arch bridge. The original arch bridge was replaced with one that could accommodate larger trucks and was also slightly longer with abutments higher above the river to make the bridge safer from flood waters. The old bridge is now a pedestrian walkway and though it is currently badly rusted, there are plans afoot to repaint the bridge. Finally, this bridge has had many names over the years, leading to some confusion at times among bridge historians. It has generally been called the Chesterfield Bridge, though contemporary news accounts also termed it the Gulf Bridge, for the large valley that leads to the bridge on the New Hampshire side. More recently, it was called the Seabee Bridge, a name that never seems to have stuck with locals, after the famed Navy fighting Seabees of World War II. It is now, since September 2010, called the Justice Harlan Fiske Stone Bridge, as signed into law by New Hampshire Governor John Lynch. It is named for Chesterfield native and noted jurist Harlan Stone, who served on the U.S. Supreme Court from 1925 until his death in 1946, serving the last five years as Chief Justice.

Appendix 1

Bridge Companies Represented in the Region

While this list is fairly complete for extant bridges (built from ca. 1869 to 1940) as of 2010, some fabricators of bridges no longer extant may have gone undocumented. Bridge companies with bridges still remaining in any of the three northern New England states show those states in parentheses.

A.D. Briggs, Springfield, Massachusetts (NH)
American Bridge Company (Ambridge) (VT, NH, and ME)
Berlin Construction Company (VT, NH, and ME)
Berlin Iron Bridge Company (VT and NH)
Bethlehem Steel Corporation (VT, NH, and ME)
Boston Bridge Works (VT, NH, and ME)
Canadian Bridge Company (ME)
Canton Bridge Company (NH)
Corrugated Metal Company
Dominion Bridge Company (ME)
Groton Bridge Company (VT, NH, and ME)
Harris Structural Steel (ME)
Henry Norton, Springfield, Massachusetts (VT)
J.R. Worcester (NH and VT)
John Hutchinson, Troy, New York
Keystone Bridge Company (Pittsburgh)
King Iron Bridge Company (VT)
Lackawanna Steel Corporation (NH and ME)
McClintik-Marshall (VT, NH, and ME)
Moseley Bridge (NH and VT)
National Bridge and Ironworks (VT)
New England Structural Company (ME)
Niagara Bridge Company
Pan American Bridge Company (NH)
Pennsylvania Steel (NH and ME)
Phoenix Bridge Company (NH, VT, and ME)
Pittsburgh Bridge Company
Pittsburgh–Des Moines Steel Company (ME)
Portland Company, Portland, Maine
R. F. Hawkins Iron Works (NH)
Union Bridge Company Athens, Pennsylvania
United Construction Company (VT, NH, and ME)
Vermont Construction Company (VT)
Wrought Iron Bridge Company

Appendix 2

Vermont Historic Metal Highway Bridges

Town	Feature Crossed	Truss type	Year Built	Location
Arlington	Batten Kill	Warren Pony	1918	Pedestrian Bridge
Barnet	Passumpsic River	Camelback Through	1928	Town Road 11
Barnet	Passumpsic River	Pratt	1928	Rte 220
Barnet	Connecticut River / Canadian Pacific RR	Camelback/ Parker	1938	Town road 6
Barre	Stevens Branch	Warren Pony	1924	Pioneer Street Pedestrian
Barton	Barton River	Warren Pony	1906	Pedestrian Bridge
Barton	Barton River	Warren Pony	1918	Lovers Lane
Berlin	Dog River	Warren Pony	1934	Rte 12
Berlin	Dog River	Pratt	1934	Rte 12
Berlin/ Middlesex	Winooski River	Parker	1928	Town Road 7
Bethel	N/A	Lenticular Pony	1896	In storage
Bethel	White River	Warren MI	1900	Town road 24
Bethel	Third Branch White River	Camelback Pony	1928	Rte 12
Bethel	White River	Parker	1928	River Street
Bethel	3rd Branch, White River	Pratt	1928	Town Road 47
Bethel	Gilead Brook	Warren Deck	1928	Rte 12
Bloomfield	Nulhegan River	Pratt	1938	Rte 102
Bradford	Waits River	Warren Pony	1934	Creamery Rd.
Brandon	Otter Creek	Warren Pony	1929	Syndicate Rd.
Bridgewater	Ottauquechee River	Warren Pony	1922	Town Rd. 34
Bridgewater	Ottauquechee River	Pratt	1928	Route 100A
Bristol	New Haven River	Camelback Pony	1925	River Street
Brookline /New Fane	West River	Camelback	1928	Town road 2
Cavendish	N/A	Pratt	1890	In storage-Old Howard Hill Bridge
Cavendish	Black River	Parker Pony	1905	Depot Street
Charlotte	Canal	Warren Pony	1925	Town Road 36
Dummerston	West River	Warren MI	1892	Rice Farm Road and Route 30
Enosburg	Tyler Branch	Warren Pony	1911	Town Road 42
Enosburg	Missisiquoi River	Parker	1929	Boston Post Rd.

Vermont Historic Metal Highway Bridges

Town	Feature Crossed	Truss type	Year Built	Location
Fairfax	Lamoille River	Parker	1930	River Road
Fairfield	Black Creek	Warren Pony	1910	Paradee Rd.
Guilford	Broad Brook	Warren Pony	1915	Town Road 17
Hardwick	Cooper Brook	Warren Pony	1915	Trail-relocated
Hartford	Railroad	Pratt Pony	1900	Gillette Street
Hartford	Quichee Gorge	Steel Deck Arch	1911	U.S. 4
Hartford	White River	Warren Deck	1928	6
Hartford	White River	Parker	1929	Town Road 4
Highgate	Mississiquoi River	Lenticular	1887	Rte
Highgate	Mississiquoi Rover	Pratt	1928	Town Road 19
Hinesburg	Canal	Warren Pony	1914	
Jamaica	West River	Pratt	1926	Town Road 19
Jamaica	Wardsboro Brook	Pratt	1936	Town Road 43
Johnson	Lamoille River	Pratt	1928	Railroad St.
Ludlow	Black River	Warren Pony	1929	Mill St.
Montpelier	Winooski River	Baltimore	1902	Granite St.
Montpelier	N. Branch Winooski River	Warren Pony	1928	Langdon St.
Montpelier	N. Branch Winooski River	Warren Pony	1928	School St.
Montpelier	Winooski River	Parker	1929	Taylor St.
Montpelier/Berlin	N/A	Pratt	1929	In Storage
Moretown	Mad River	Warren Pony	1920	Town Rd 39
Morristown	Lamoille River	Pratt	1926	Bridge St.
Morristown	Lamoille River	Pratt	1928	Cady's Falls Rd.
Morristown	Lamoille River	Pratt	1928	Cady's Falls Rd.
New Haven /Weybridge	Otter Creek	Warren	1908	Pearson Rd.
Newfane	Rock River	Pratt	1929	Parish Rd.
Northfield	Pedestrian, over NECRR	Parker Patent	1870	Vine St.
Northfield	Dog River-Over RR	Parker Pony	1908	Rabbitt Hollow Rd.
Northfield	Dog River	Warren	1924	Pedestrian Bridge
Orleans	Barton River	Tied Arch w/Truss	UNK.	Pedestrian Bridge
Poultney	Poultney River	Pratt	1923	Rte 31
Poultney	Poultney River	Parker Pony	1925	Thrall Rd.
Putney	NEC Railroad	Pratt Pony	UNK.	Town Rd. 52
Richford	Mississiquoi River	Parker Pony	1934	Main St.
Richford Canada	Mississiquoi River	Parker	1929	Rte 105
Richmond	Winooski River	Parker	1928	Bridge St.
Richmond	Winooski River	Pennsylvania	1929	U.S. 2
Rockingham	Williams River	Warren Deck	1929	U.S. 5 Royalston
Royalton	White River	Camelback	1928	Royalston Hill Rd.
Royalton	White River	Parker	1928	Town Rd. 36
Rutland	Otter Creek	Pratt Through w/Pony	1928	River St.
Rutland	Otter Creek	Warren Pony	1928	Ripley Rd.
Sharon	White River	Parker	1928	River Rd.

Town	Feature Crossed	Truss type	Year Built	Location
Sheldon	Black Creek	Warren Pony	1903	Bouchard Rd.
Sheldon	Missisiquoi River	Parker	1928	Shawville Rd.
Sheldon	Missisiquoi River	Parker	1928	N.Sheldon Rd.
Springfield	Black River	Baltimore	1929	Paddock Rd.
Stockbridge	White River	Pratt	1929	Rte 100
Stockbridge	White River	Warren	1929	Town Rd. 6
Swanton	Missisiquoi River	Pennsylvania	1902	Pedestrian Bridge
Thetford	East Ompompan-oosuc Branch	Warren Pony	1910	Pedestrian Bridge
Tunbridge	First Branch White River	Warren Pony	1889	Foundry Rd.
Wallingford	Otter Creek	Warren Pony/Camelback		Pedestrian Bridge
Waterbury/	Little River	Parker	1928	Winooski St.
West Rutland	Clarendon Brook	Warren Pony	1915	Pedestrian Bridge
Westfield	Mill Brook	Warren Pony	1910	Trail relocated
Westhaven/ New York	Poultney River	Warren Pony	1921	Brook Rd.
Wilmington	Deerfield River	Warren DI	1896	Bypassed
Wolcott	Wild Branch, Lamoille River	Camelback Pony	1928	Town Rd. 2
Wolcott	Lamoille River	Warren Pony	1928	Town rd. 3
Woodstock	Ottauquechee River	Parker Patent	1870	Elm St.
Woodstock	Ottauquechee River	Pennsylvania Modified w/ Steel Arch	1900	Mill St.

Appendix 3

New Hampshire Historic Metal Highway Bridges

Town	Feature Crossed	Truss type	Year Built	Location
Antrim	Contoocook	Pratt	1893	Thompson Crossing Rd.
Ashland	Pemigewasset	Pratt Steel Deck	1931	US Rte 3 to Bridgewater
Bartlett	Rocky Branch	Pratt	1936	US Rte 302
Bath	Ammonoosuc	Warren	1928	US Rte 302
Berlin	Androscoggin	Steel Combination	1916	Bridge St.
Bethlehem	Ammonoosuc	Warren	1927	NH Rte 142
Bethlehem	Ammonoosuc	Pratt	1928	US Rte 302
Bethlehem	Ammonoosuc	Warren	1928	Prospect Street
Boscawen	Merrimack	Parker	1907	Depot Street
Campton	Pemigewasset	Lenticular deck	1886	Livermore Falls–off Rte 3
Campton	Beebe River	Warren pony	1920	Eastern Corner Rd.
Canaan	Indian River	Warren pony	1913	Lary Road
Carroll	Ammonoosuc	Warren pony	1915	River Road
Charlestown	Connecticut	Pennsylvania	1930	To Springfield, VT
Chesterfield	Connecticut	Steel Arch	1936	Rte 9 to Brattleboro, VT
Chichester	Suncook	Lenticular	1887	Depot Street
Claremont	Sugar	Bowstring Arch	1870	Footbridge Monadnock Mill #6
Concord	Merrimack	Pratt	1915	Sewall's Falls Rd.
Dalton	Connecticut	Pratt Steel Deck	1928	Gilman Rd/Whitcomb Bridge to Lunenberg, VT
Danbury	Smith River	Warren pony	1913	Pleasant Road
Deering	Contoocook	Warren pony	1905	West Deering Rd. to Antrim
Dover	Cocheco	Warren pony	pre–1895	Pedestrian-Cocheco Mills
Dover	Cocheco	Warren pony D I	pre–1925	Watson Rd. at the falls
Dover	Piscataqua	Continuous Steel	1934	General Sullivan Bridge
Effingham	Ossipee	Pratt	1936	Huntress Bridge Rd. to Freedom
Exeter	Railroad	Warren pony	1892	Park St. over BM Railroad

Town	Feature Crossed	Truss type	Year Built	Location
Franconia	Ham/Gale	Lenticular pony	1889	Rte 18, former Delage Rd.
Franconia	Gale	Lenticular pony	Unk	Dow Avenue
Greenville	Souhegan	Pratt	1938	Power Dam Bridge
Hancock	Contoocook	Pratt pony	1905	Cavender Rd. to Greenfield
Haverhill	Connecticut	Baltimore truss	1903	Railroad over auto to Wells River, VT
Haverhill	Connecticut	Steel Arch	1923	To Newbury, VT
Hebron	Cockermouth	Warren pony	1921	Braley Road
Henniker	Contoocook	Pratt	1915	Patterson Hill Rd.
Henniker	Contoocook	Pratt	1933	Western Ave.–Moved from Concord, originally built 1915
Henniker	Contoocook	Warren	1937	Ramsdell Rd.
Hinsdale	Connecticut	Pennsylvania	1920	NH Rte 119 to Hinsdale/Brattleboro, VT
Hinsdale	Connecticut	Parker	1926	Rte 9
Hooksett	Merrimack	Pratt	1909	Off Rte 3A–Lilac Bridge
Jackson	Wildcat Brook	Parker pony	1905	Chef Rd.
Lebanon	Connecticut	Steel Combination	1936	US Rte 4
Littleton	Ammonoosuc	Pratt	1928	Bridge St.
Littleton	Ammonoosuc	Pratt	1928	Reddington St./Apthorp Br.
Lyme	Connecticut	Parker	1937	East Thetford Rd. to VT
Manchester	Merrimack	Warren Steel Deck	1923	Queen City Ave.
Milford	Souhegan	Suspension	1889	Bridge St.
Milford	Souhegan	Pratt	1910	Jones Crossing
Monroe	Connecticut	Parker	1930	McIndoes Rd. to Barnet, VT
Monroe	Connecticut	Parker	1937	North Monroe
Nashua	Nashua River	Pratt	1903	Nashua Mfr. Co.–Clocktower Place
Newport	Sugar	Warren	1937	Pine Hill Rd./Oak St.
Northumberland	Ammonoosuc	Pratt	1939	US Rte 3
Orford	Connecticut	Steel tied Arch	1938	To Fairlee, VT–Samuel Morey Memorial Bridge
Piermont	Connecticut	Pennsylvania	1928	NH Rte 25 to Bradford, VT
Plymouth	Baker	Pratt	1930	US Rte 3
Portsmouth	Piscataqua	Vertical Lift	1921	US Rte 1 to Kittery, ME
Portsmouth	Piscataqua	Vertical Lift–double deck	1940	Rte 1 Bypass
Rollinsford	Railroad	Warren pony	1901	Oak St.
Salisbury	Blackwater	Pratt pony	1893	moved to private location

Town	Feature Crossed	Truss type	Year Built	Location
Shelburne	Androscoggin	Combination	1897	North Rd./Meadows Br.
Stark	Ammonoosuc	Warren pony	1909	Bell Hill Rd.
Stark	Ammonoosuc	Warren pony	1920	Paris Road
Stewartstown	Connecticut	Steel Arch Deck	1930	Beecher Falls Road
Stratford	Connecticut	Pratt	1893	Stratford Hollow to Maidstone, VT
Sugar Hill	Gale	Warren	1928	Jesseman Road
Tilton	Winnepesaukee	Truesdell lattice	1881	To Tilton Island
Troy	Private drive	Low Warren?	ca. 1910	Marlborough Rd.– originally on Shaker Brook
Weare	Piscataquog	Warren pony	1940	Rockland Bridge
Wentworth	Baker	Warren	1909	Wentworth Village Rd.
Wentworth	Baker	Warren	1930	Sanders Hill Rd.

Appendix 4

Maine Historic Metal Bridges

Town	Feature Crossed	Truss type	Year Built	Location
Alna	Sheepscot River	Warren pony	1936	Dock Rd.
Andover	W. Branch Ellis River	Pratt pony	1935	Rte 120
Arrowsic	Sassanoa River	Cantilver arch	1950	Rte 127
Auburn	Little Androscoggin River	Pratt	1937	Hotel Rd.
Auburn	Androscoggin River	Warren	1937	BroadStreet
Augusta	Kennebec River	Cantilver deck arch	1950	Near old Fort Western
*Bath	Kennebec River	Warren truss vertical lift span tracks	1926	Maine Eastern RR next to U.S. 1
*Biddeford	Elm Street	Baltimore	1928	Elm Street
Boothbay	Back River	Warren pony swing span	1931	Barters Island
Brownfield	Saco River	Warren	1936	Rte 160
Brownville	Pleasant River	Warren	1935	Rte 11
Brunswick	Androscoggin River	Baltimore truss railroad with suspended lower deck	1909	Off U.S. 1
Brunswick	Androscoggin River	Warren	1932	U.S. 201
Brunswick	Androsoggin	Suspension	1892/1936	Off U.S. 1
Buxton	Saco River	Warren	1937	Rte 112
Chesterville	Wilson Stream	Camelback	1937	Rte 156
Clinton	Sebasticook River	Warren	1936	Gogan Rd.
Deer Isle	Eggemoggin Reach	Suspension	1931	Rte 15–Deer Isle–Sedgewick
Detroit	East Branch Sebasticook River	Warren pony	1936	Rtes 69 & 220
Dresden	Middle River	Warren	1936	Rte 197
Durham	Androscoggin River	Continuous Warren	1937	Rtes 9 and 125
East Machias	East Machias River	Warren pony	1937	Rte 191
Edmunds Twp	Crane Mill Stream	Parker	1936	U.S. 1
Falmouth	Presumpscot River	Warren deck	1932	Rtes 26 and 100
Fort Kent	St. John River	Pennsylvania	1929	U.S. Rte 1 and Rts. 11 & 161
Gilead	Androscoggin River	Camelback	1921	North Rd.
Greene	Androscoggin River	Warren	1937	Center Bridge Rd.
Harrison	Crooked River	Double-Intersection Warren	1912	Bow Street

Town	Feature Crossed	Truss type	Year Built	Location
Hollis	Androscoggin River	Continuous Warren	1937	Rte 4A
Hollis	Canal-Androscoggin River	Warren	1937	Rte 4A
Howland	Piscataquis River	Pennsylvania	1929	Coffin St.–Scheduled for removal 2010–11
Kenduskeag	Kenduskeag Stream	Pratt	1932	Town House Rd.
Kittery	Piscataqua River	Vertical lift Warren	1921	U.S. 1
Kittery	Piscataqua River	Vertical lift Warren	1940	U.S. 1 Bypass
Leeds	Dead River	Pennsylvania	1922	Rte 106
Limington	Little Ossipee River	Warren pony	1923	Nason Mills Rd.
Limington	Saco River	Warren	1937	Rte 11-Steep Falls
Madawaska	St. John River	Pennsylvania	1921	Rte. 50 spur
Mexico–Rumford	Androscoggin River	Warren	1931	New Page Corp bridge
Milford	Otter Stream	Warren	1936	County Rd.
Milo	Pleasant River	Warren	1936	Pleasant Street
Naples	Crooked River	Warren pony	1937	State Park Rd.
New Portland	Carrabasset River	Suspension	1866	Wire Bridge Road
New Sharon	Sandy River	Pennsylvania	1916	Adjacent U.S. 2
Old Town	Pushaw Stream	Warren pony	1937	Rte 16
Passadumkeag	Passadumkeag Stream	Warren pony	1937	Goulds Ridge Rd.
Peru	Androscoggin River	Parker	1930	N. Main Street
Pittsfield	N/A	Kingspost	1894	Maine DOT facility
*Portland	St. John St	Baltimore	1890	St. John Street
Prospect-Verona	Penobscot River	Suspension	1931	Closed–old U.S. 1
Richmond	Kennebec River	Parker swing span	1931	Rte 197
Rumford	Androscoggin River	Steel arch	1935	Morse Street
Southport	Townsend Gut	Warren swing span	1939	Rte 27
Thomaston	St. George River	Pratt (former bascule span)	1925	River Road
Waterville	Kennebec River	Suspension	1903	Footbridge
Windham	Presumpscot River	Warren pony	1912	Gambo falls

* denotes railroad bridge

Bibliography

Published Works

Allen, Richard Sanders. *Covered Bridges of the Northeast*. Brattleboro, VT: Stephen Greene Press, 1957.

Armstrong, John B. *Factory Under the Elms: A History of Harrisville, New Hampshire, 1774–1969*. N. Andover, MA: Museum of American Textile History, 1985.

Atwood, R. E. *Stories and Pictures of the Vermont Flood November, 1927*. Burlington, VT: Atwood, 1927.

Condit, Carl W. *American Building Art: The Nineteenth Century*. New York: Oxford University Press, 1960.

Conwill, Joseph D. *Maine's Covered Bridges*. Charleston, SC: Arcadia, 2003.

Dartmouth College. *A Catalogue of the Officers and Students of Dartmouth College for the Academical Year 1852–53*. Hanover, NH: Dartmouth Press, October 1852.

Deetz, James. *In Small Things Forgotten: The Archaeology of Early American Life*. New York: Doubleday, 1977.

Delorme Mapping Company. *The Maine Atlas and Gazetteer*. Freeport, ME: Delorme, 1988.

____. *The New Hampshire Atlas and Gazetteer*. Freeport, ME: Delorme, 1988

____. *The Vermont Atlas and Gazetteer*. Freeport, ME: Delorme, 1988.

Edwards, Llewellyn Nathaniel. *A Record of History and Evolution of Early American Bridges*. Orono, ME: University Press, 1959.

Farnham, Euclid. *Tunbridge Past*. Tunbridge, VT: 1971.

Foss, Roland E. *A History of the New Portlands in Maine*. Farmington, ME: Knowlton and McLeary, 1977.

Gale Group. *Notable Twentieth-Century Scientists*. Detroit: Gale Group, 1995.

Garvin, Donna-Belle, and James L. Garvin. *On The Road North of Boston: New Hampshire Taverns and Turnpikes 1700–1900*. Hanover, NH: University Press of New England, 1988.

Gies, Joseph. *Bridges and Men*. Garden City, NY: Doubleday, 1963.

Hancock Historical Commission. *The Second Hundred Years of Hancock, New Hampshire*. Canaan, NH: Phoenix, 1979.

Harlow, Alvin F. *Steelways of New England*. New York: Creative Age Press, 1946.

Hool, George A., and W. S. Kinne. Revised by R. R. Zipprodt and H. E. Langley. *Moveable and Long-Span Steel Bridges*. New York: McGraw-Hill, 1943.

Johnson, Clifton. *Highways and Byways of New England*. New York: MacMillan, 1915.

Karr, Ronald Dale. *Lost Railroads of New England*. Pepperell, MA: Branch Line Press, 1989.

Knoblock, Glenn A. *New Hampshire Covered Bridges*. Charleston, SC: Arcadia, 2002.

Maine, State of. *Railroad Commissioner's Report*. Augusta: 1892.

____. *Sixth Annual Report Commissioner of Highways*. Augusta: Kennebec Journal Print, 1910.

McCullough, Robert. *Crossings: A History of Vermont Bridges*. Barre: Vermont Historical Society, 2005.

Merriman, Mansfield, and Henry S. Jacoby. *A Text-Book on Roofs and Bridges, Part I*. New York: John Wiley, 1913.

_____, and _____. *A Text-Book on Roofs and Bridges, Part III*. New York: John Wiley, 1909.
Metcalf, Henry, and John Norris. *The Granite State Monthly XXXVI*. Concord, NH: 1904.
Milan, New Hampshire. *Historical Notes of Milan, New Hampshire*. Littleton: Courier Printing, 1971.
New Hampshire, State of. *40th Annual Report of the Railroad Commissioners*. Concord: Parsons B. Cogswell, 1884.
Parker, Edward, ed. *History of Nashua, New Hampshire*. Nashua: Telegraph Publishing, 1897.
Ratigan, William. *Highways over Broad Waters: Life and Times of David B. Steinman, Bridgebuilder*. Grand Rapids, MI: Wm. B. Eerdmans, 1959.
Sloane, Eric. *Our Vanishing Landscape*. New York: Wilfred Funk, 1955.
Steinman, David B., and Sara Ruth Watson. *Bridges and Their Builders*. New York: Dover, 1957.
Storrs, John W., and Edward D. *A Handbook for the Use of Those Interested in the Construction of Short Span Bridges*. Concord, NH: Storrs, 1918.
Varney, Geo. J. *A Gazetteer of the State of Maine*. Boston: B.B. Russell, 1886.
Vermont, State of. *10th Biennial Report of the Board of Railroad Commissioners*. Bradford: Opinion Press, 1906.
_____. *13th Biennial Report of the Public Service Commissioners, 1910–1912*. Bellows Falls: Gobie Press, 1912.
Vose, George W. *Bridge Disasters in America: The Cause and Remedy*. Boston: Lee and Shepard, 1887.
Webster, Kimble. *History of Hudson, New Hampshire*. Manchester: Granite State, 1913.
Whitcher, William. *History of the Town of Haverhill, New Hampshire*. Concord: Rumford Press, 1919.
Whittlesey, Charles W. *Crossing and Re-Crossing the Connecticut River*. New Haven: Tuttle, Morehouse, & Taylor, 1938.

Newspaper, Magazine, and Journal Articles

Allen, Richard Sanders. "The Bridge That Spans a Century," *Steelways* 57, no 5 (October 1957): 24.
Ashland, NH. "A Steel Bridge," *Laconia Democrat*, April 18, 1902.
Associated Press. "Some Bemoan Proposed Design of NH-Maine Bridge," *Conway Daily Sun*, December 4, 2010, 7.
"At Least Fifty Killed by the Wrecking of a Montreal Express." *The New York Times*, February 6, 1887.
Bethlehem, NH. "Bethlehem Has Three Iron Bridges Go Out," *The Courier*, November 6, 1927.
Concord, NH. "But Sewalls Falls Bridge Renaming Is Premature," *Concord Monitor*, January 28, 2009.
Conley, Casey. "Making Tracks: Groups Push for Revival of Rail Service in Maine," *Conway Daily Sun*, September 18, 2010, 18.
Conway Scenic Railroad. "The History of the Frankenstein Trestle," June 2007.
Cunningham, Geoff Jr. "Bridge Money Delivered," *Fosters Daily Democrat*, October 21, 2010, 1.
_____. "Hybrid Bridge Weighed," *Fosters Daily Democrat*, June 23, 2010, 1.
Delony, Eric. "Wrought Iron and Steel Bridges," *The Journal of the Society for Industrial Archeology* 19, no. 2 (1993).
Garvin, James L. "High Water: Rebuilding Bridges after the Floods of 1927 and 1936," *New Hampshire Highways*, March/April 2004.
Grossman, Max R. "'Most Bitterly Hated,' Says New England's Oldest Mayor," *Boston Sunday Post*, November 9, 1941, A6.
Harrigan, John. "Sporadic $1 Toll Days Keep a Bridge from Falling Down," Manchester (NH) *Sunday News*, June 15, 1975 (clipping, no page).
Hartgen, David T., Ravi K. Karanam, M. Gregory Fields, and Adrian T. Moore. "18th Annual Report on the Performance of State Highway Systems (1984–2007/8)," The Reason Foundation, Washington, DC, 2009.

Highgate, VT. "Highgate Bridge Emerges From Rust, Ceremony Wednesday," *The St. Albans Messenger*, October 10, 2000.
Vose, George L. "Bridge Inspection in Maine," *The New York Times*, February 25, 1878.
Weingardt, Richard G. "John Alexander Low Waddell: Genius of Moveable Bridges," *Structure Magazine*, February 2007, 61–64.

Internet Sources

American Society of Civil Engineers. "Report Card for American Infrastructure — Maine," http://www.infrastructurereportcard.org/state-page/maine.
_____. "Report Card for American Infrastructure — New Hampshire," http://www.infrastructurereportcard.org/state-page/new-hampshire.
_____. "Report Card for American Infrastructure — Vermont," http://www.infrastructurereportcard.org/state-page/vermont.
Baughn, James. "Historic Bridges of the United States," http://bridgehunter.com.
Boothby, Thomas. "Designing American Lenticular Truss Bridges 1878–1900," http://www.historycooperative.org/cgi-bin/printpage.cgi.
Brattleboro Community Brain Trust. "Story of the West River Railroad," http://www.ibrattleboro.com/braintrust/index.php?title=Story_of_the_West_River_Railroad.
Brown, Kathi. "Moveable Bridge Hall of Fame Page," http://heavymovablestructures.org/other/hall_of_fame/pdf/Harrington.PDF.
Brunswick-Topsham Swinging Bridge. "Frequently Asked Questions," http://www.saveourbridge.org/pages/faq.html.
Connors, Frank. "The Bridge Over the Cathance," http://www.link75.org/mmb/History/cathbr.html.
Cooper, Theodore. "American Railroad Bridges — Engineering News — July 6, 1889," http://www.catskillarchive.com/rrextra/brcoo0.html.
Delony, Eric. "The Golden Age of the Iron Bridge" http://americanheritage.com/articles/magazine/it/1994/2/1994_2_8.shtml.
Garvin, James L. "Shelburne Meadows Bridge," http://www.shelburnenh.com/bridge.html.
Hall, B. A., Durham Oral Histories, http://durhamme.com/durham-life.html.
Hanson, Eric. "National Register Nomination Information — Rice Farm Road Bridge," http://www.crjc.org/heritage/VO3–18.htm.
Higgins, Pat. "Explosive Times at Vanceboro," http://imaginemaine.com/Vanceboro.html.
Hill, Cindy Ellen. "Iron Beauties: Vermont's Metal Truss Bridges Span History and Carry Us Forward," http://www.livinmagazine.com/2010/01/04/iron-beauties/.
Lehigh University Digital Library. "Illustrated Pamphlet of Wrought Iron Bridges Built by the Wrought Iron Bridge Company, Canton, Ohio," http://digital.lib.lehigh.edu/cdm4/bridges_viewer.php.
Maine Department of Transportation. "History of Railroading in Maine," http://www.maine.gov/mdot/freight/railroading-history.php.
Maine Section, American Society of Civil Engineers. "Maine Historic Civil Landmarks," http://www.maineasce.org/downloads/History_Heritage/MeHistCivLandmarks6–05.pdf.
Miller, M. Nadine. "National Register Properties — Quechee Gorge Bridge," http://www.crjc.org/heritage/V11–30.htm.
Mohney, Kirk F. "Daniel Beedy 1810–1889 Maine Architect," http://homepages.rootsweb.ancestry.com/~nvjack/beede/daniel_beedy.htm.
Nashua River Watershed Association. "Recreational Gems Throughout the Watershed," http://www.nashuariverwatershed.org/recreation.html.
National Trust for Historic Preservation. "11 Most Endangered Historic Places — Memorial Bridge," http://www.preservationnation.org/travel-and-sites/sites/northeast-region/memorial-bridge.
Parsons, Graham. "The Central Vermont Railway: An Essay. http://www.images.technomuses.ca/?en/stories/central_vermont/b/page/1.
Petroski, Henry. "Waldo-Hancock Bridge," http://www.britannica.com/bps/additionalcontent/18/22859130/WaldoHancock-Bridge.

Rochester (Vermont) Historical Society. "The White River Valley Railroad," *http://www.rochester-historical.org/?page_is=131.*

Archival Sources

Boston and Maine Railroad, ICC Valuation Survey, 1914, held at Boston and Maine Railroad Historical Society, Center for Lowell History, University of Massachusetts at Lowell.

Bridge Records, Maine Central Mountain Division, held by the Conway Scenic Railroad, Conway, NH.

Maine Central Railroad Annual Reports, 1870–1900, held at Boston and Maine Railroad Historical Society, UMass, Lowell.

Morse, Viola Bishop. "School Days," recollections written ca. 1980s, Medburyville, VT, held in file cabinet, School District #7.

Historic American Engineering Record (HAER) Reports

"Manchester Street Bridge, Concord, NH," No. NH-28
"Notre Dame Bridge, Manchester, NH," No. NH-14
"Osgood Bridge, Campton, NH," No. NH-10

Maine Department of Transportation Records

"Highway Bridge Building in Maine by the Maine State Highway Commission, 1905–1956," 1997.
Historic Bridge Inventory Forms, 1997 Survey All Pre–1950 Maine Bridges.
"2008–2011 STIP Amendments for ARRA Projects (Stimulus)

New Hampshire Historic Property Documentation Reports

Bartlett Bridge, No. 191/139 (Rocky Branch Bridge) — Richard M. Casella
Central Street Bridge (Bristol), No. 113/064 — James L. Garvin
Chesterfield-Brattleboro Arch Bridge, No. 040/095 — Richard M. Casella
Meadow Bridge (Shelburne), No. 122/110 — Richard M. Casella
Newfields-Stratham Swing Bridge, No. 132/066 — Richard M. Casella
North Stratford-Bloomfield Bridge, No. 029/206 — Richard M. Casella
Samuel Morey Memorial Bridge (Orford-Fairlee), No. 062/124 — Richard M. Casella
Stratford-Maidstone Bridge, No. 098/064 — Richard M. Casella and Stuart P. Dixon

New Hampshire Division of Historical Resources Records

Garvin, James L. "Connecticut River Bridges," 2004
_____. "Iron and Steel Truss Bridges in New Hampshire," September 2008.
_____. "National Register of Historic Places Registration Form-Memorial Bridge," draft copy, December 2008.
_____. "Position Paper: Preservation and Reuse of Historic Bridges As Economic Stimulus," December 2008.
Kenison, Leon S. (Commissioner, NH DOT) Letter to Carl W. Schmidt, Orford Historical Society, regarding color scheme of Samuel Morey Bridge, dated August 20, 1999.
New Hampshire. "Highway Department Standard Specifications for Road and Bridge Construction, May 1, 1935."
Zoller, Jerry S. (NH Bureau of Bridge Design) Inter-Department Communication to Robert Landry (Project Engineer) regarding color of Samuel Morey Bridge, March 15, 2001.

Vermont Agency of Transportation Records

"A Brief Study of Vermont's Metal Truss Bridges," n.d.
"FHWA ARRA Projects Master List," March 25, 2010 (Stimulus).

Bibliography

Gillies, Paul. "The History and Law of Vermont Town Roads," October 28, 2006.
McCullough, Robert, and Susan Scribner. "Vermont Historic Bridge Program — Bridges to Be Adapted to Alternative Transportation Uses," April 1999.
"Project Status Report-BHF 6400(31), Taylor Street Bridge, Montpelier," VAOT.
"State Owned Railroad Bridges," Rail division of VAOT.
"Vermont Historic Metal Truss Bridge Study — Bridges Listed by County," June 13, 1997.
"Vermont Transportation Economic Stimulus and Recovery — Bridges."

Annual Town Reports

Bethlehem, NH 1922, 1928, 1929
Boscawen, NH 1907, 1908
Concord, NH 1874, 1875, 1898, 1915
Dalton, NH 1928, 1929
Dover, NH 1890–1935
Henniker, NH 1901, 1916
Highgate, VT 1892
Monroe, NH 1930

Personal Interviews

Mr. Paul Hallett, Operations Manager, Conway Scenic Railroad, Conway, NH, July 30, 2010.
Mr. Duane Lewis, Bridge Operator, Southport, Maine, April 24, 2010.
Mr. Robert McCullough, Vermont Historic Bridge Program, Montpelier, VT, May 21, 2010.

Index

Numbers in ***bold italics*** indicate pages with photographs.

academic structures 30
Addison, VT 27
Alburg, NY 44, 60
Alexander Scammell Bridge (NH) 44
Alna, ME 23
Ambassador Bridge (MI-CAN) 166
American Bridge Company (Ambridge) 55, 58, 95, 97, 113, 119, 122, 124, 129, 152, 157, 173, 176, 182
American Institute of Steel Construction 173, 182
American Jobs Act 85
American Reinvestment and Recovery Act *see* Stimulus
American Society of Civil Engineers (Maine Chapter) 40, 165
Ammonoosuc River 137
Amtrak *Downeaster* 83
Amtrak *Ethan Allen Express* 83
Amtrak *Vermonter* 103
Anderson, John 135
Anderson, Samuel 136
Andrews, D.H. 141
Androscoggin Railroad 145
Androscoggin River 138,142, 144, 160, 161
Antrim, NH 19, 33, 35, 82, 115, 116
Arlington, VT 31
Armour, Swift, and Burlington Bridge 45
Aroostook River Bridge (ME) 39, ***69***
Arrowsic, ME 39, 57, 148
Art Deco style ***42***
Art Moderne style 172
Ash Street Bridge (NH) 60
Ashland, ME 69
Ashland, NH 21, 70, 125, 126
Ashland-Bridgewater Bridge (NH) 70
Ashland Paper Company 126
Ashuelot Railroad Bridge (NH) 68
Athens, PA 136
Atlantic & St. Lawrence Railroad 64, 67
Auburn, ME 75
Augusta, ME 39, 57, 65, 75, 145
Augusta Memorial Bridge 39

Babbitt, Charles 60, 61
Babbitt, Edward 60
Bailey Bridge, ME 81
Baker River 128
Ballard, Ephraim, Jr. 7
Baltimore & Ohio Railroad 10
Baltimore Bridge Company 36
Baltimore truss 36, 93, 94, 103, 122, 132, 140, 144, 178
Bancroft, ME 66
Bangor, ME 56
Bangor & Aroostook Railroad 67, 69, 70, 75
Barbour, Volney 61
Barnet, VT 69, 74, 176
Barre, VT 21, 86
Bars Mills Bridge (ME) 38
Bartlett, NH 28, 30, 36, 60, 132, 133
Barton, VT 86
Bath, ME 46, 63, 146–147
Bath, NH 75
Bay Bridge (NJ) 45
Bedell, Moody 6
Beecher Falls Bridge (NH) 59, ***172***–173
Beedy, Daniel 164
Benjamin, Asher 7
Bennington, NH 19, 33, 81, 82, 115, 116
Bennington, VT 32, 121
Bennington Museum 81
Berlin, NH 13, 35
Berlin, VT 74
Berlin Construction Company 17, 53, 123, 176
Berlin Iron Bridge Company 28, 29, 34–35, 37, 40, 48, 58, 91, 106, 111, 116, 117, 128, 131, 174, 182, 184
Bessemer process 12
Bethel, VT 24, 35, 68, 86, 100, 101
Bethlehem, NH 75
Bethlehem Steel Company 118, 146, 184
Biddeford, ME 32, 36, 140
Bingham, ME 35, 65, 66
Bishop, George 61
Black River 93, 94
Blacksmith Shop Bridge (VT) 80
Bloomfield, VT 35, 174
Board of Highway Commissioners (VT) 18

Bog Brook Bridge (NH) 61
Borough Bridge (NH) 53
Boscawen, NH 37, 82, 124
Boston & Maine *Alouette Express* 179
Boston & Maine Railroad 36, 56, 58, 60, 64, 67, 68, 71, 97, 113, 114, 120, 125, 126, 170, 179, 182
Boston & Maine *Red Wing Express* 179
Boston Bridge Works Company 56, 63, 66, 67, 69, 137, 141, 154, 177, 180
Boston, Concord, & Montreal Railroad 179
Bowdoinham, ME 33, 34
bowstring arch truss 32, 121
Boynton, Ray 157
Bradford, VT 75, 180
Brandon, VT 74
Brattleboro, VT 39, 40, 77, 82, 183, 184
Brattleboro & Whitehall Railroad 68
Bretton Woods, NH 134, 137
Bridge Act of 1915 (ME) 17
Bridge Street Bridge (NH) 59
Bridge Street Bridge (VT) 85
bridges: abutments 23; alterations 82–84; approaches 24; bascule bridges 43–44, 153; builder's plaques 30; bypassed 82; cast iron 11; catalog bridge 48; chords 19; construction crews 51; cost overruns 49; counter diagonals 20; dead load 22; deck type 22; diagonals 19; disposal of 54; factory bridges 122, 158; floor beams 23; *friction rollers* 24; functionally obsolete 79; half-through truss 22; hinged-arch bridges 41, 161, 172, 177; life expectancy of 79; literature 1; live load 22; load limits 22; metaphors 1; mill complexes, use in 13–14; moveable 43–46, 146, 148; opening ceremonies 51–52; paint removal 79–80; paint schemes 25–27; piers 24; pin connected 19, 113, 114, 115, 138, 161, 178, 183; plate girder bridges 64–65; preservation

203

Index

funding 83; preservation options 80–83; rail trail, use on 85; railroad bridges 64–71; red list 83–84; reuse of 54; riveted 19, 21; salesmen 48; skew 23, 120, 140; songs 1; spandrel arch 96; spans 24; steel arch bridges 41–43, 97, 181, 183; stress testing 20; stringers 23; suspension bridges 39–40; sway bracing 22; swing bridges 43–45,148, 151; television advertising, in 78; through truss 22; tied-arch bridges 41; trestle bridges 65; vertical lift bridges 43–46, 146, 167, 170; watch list *see* red list; web members 19
Bridgewater, NH 21
Bridgewater, VT 74
Briggs, Albert D. 130
Briggs, John C. 8
Bristol, NH 75
Broadway Street Bridge (NH) 23, **120**
Brooklyn Bridge (NY) 12, 40, 63, 143
Brown, Benjamin 7
Brown, F.W. 60
Brownfield, ME 75
Brunel, Isambard K. 35
Brunswick, ME 33, 37, 71, 142–145
Bucksport, ME 157
Bulkeley Bridge (CT) 159
Bureau of Public Roads 15, 57, 75, 77
Burlingame, R.S. 182, 184
Burlington, VT 61
Buxton, ME 35, 38, 75
Buzzards Bay Bridge (MA) 171

Cabot Cotton Mill 143
camelback truss 11
Campton, NH 16, 19, 27, 35, 36, 60–61, 72–73, 126–127
Canaan, VT 60, 172
Canadian Bridge Company 166
Canadian Pacific Railroad 179
Canfield, August 10
Canterbury, NH 82, 124
cantilever truss 38–39, 148
Canton Bridge Company 48, 113
Caribou, ME 39
Carlton, Frank 146
Carlton Bridge (ME) 46, 55, 63, 146–**147**
Carrabasset River 162, 163
Carquinez Strait Bridge (CA) 63
Carroll, NH 69
Carter, Mace 149
Cavender Road Bridge (NH) 82, 114–**115**
Cavendish, VT 19, 33, 37, 86, 94
Center Rutland Bridge (VT) **74**
Central Avenue Bridge (NH) 25
Central Vermont Railroad 37, 64, 68, 74
Charlestown, NH 36, 94, 182, 183
Checkerboard House Bridge (VT) 104–**105**

Chef Road Bridge (NH) 37, 83
Cheshire Bridge (NH-VT) 182–**183**
Chesterfield, NH 40, 59, 183, 184
Chesterfield-Brattleboro Bridge (NH-VT) 40, 82, **183**, 184
Chichester, NH 25, 28, 29, 35, 82, 86, 116
Chief Justice Harlan Fiske Stone Bridge (NH-VT) 183–**185**
Childs, Enoch 8, 59
Childs, Horace 8, 59
Childs, John W. 59
Childs, Warren 8, 59
Claremont, NH 13, 26, 32, 33, 36, 65, 70, 121, 122
Clarendon, VT 86
Clarendon & Pittsford Railroad 83
Clark, "Boston" John 8
Clark, Capt. Charles 163
Clark, Charles H. 60
Clinton, ME 35
Cole's Carriage Shop 128
Columbia Bridge (NH) 17
Comerford Dam 177
Concord, NH 16, 24, 52–54, 79, 123–124, 126
Concord Railroad 58, 67
Connecticut River 172, 173, 175, 176, 177, 178, 179, 180, 181, 182, 183, 184
Connecticut River Development Corporation 176
Connecticut River Joint Commission 175
Connecticut River Toll Bridge (NH-VT) **179**
continuous truss 38, 117, 145, 146
Contoocook River 114
Conway, NH 36, 68, 134
Conway Scenic Railroad 68, 70–71, 134, 135, 136, 137
Corrugated Metal Corporation 14, 34, 36
Crane Hill Road Bridge 80
Crawford Notch, NH 65,135, 136
Crooked River 159, 160
Cuttingsville Trestle (VT) 65, 70, **93**

Dalton, NH 75, 175
Damon, Captain Isaac 8
Daniels, Secretary of War Josephus 169
Dartmouth College 61
Davis, E.B. 60
Davis, Robinson 8
Day, Fred 174
Deer Isle-Sedgewick Bridge (ME) 40, 55, 63, 85, 154–**155**, 156–157
Deerfield River 89
Delafield, Captain Richard 10
Delage, Albert 130
Delage Bridge (NH) 35, 130–**131**
Delaware & Hudson Railroad 85
Delaware Aqueduct (NJ-PA) 163
Dennis Bridge (NH) 25, 114
Depot Bridge (NH) 35
Depot Street Bridge (NH) 37

Depot Street Bridge (VT) 94–**95**
determinate structures 20
Detroit, ME 75
Detroit Graphite Company 25
Devine, M.M. 25
Dewey, A.G. 97
Dewey's Mills, VT 68
Dewey's Mills Bridge (VT) 97
Diamond Prospecting Company 122
Dixon, IL 130
Dock Road Bridge (ME) **23**
Dog River 102
Dole Bridge (NH) 61
Douglas, William 35
Dover, NH 13, 23, 25, 60, 67, 80, 81, 120
Dover Point, NH 38, 44, 117
Dover Point-Newington Bridge (NH) 117
Dow Avenue Bridge (NH) 35, 80, 131–**132**
Draffin, Jasper 61
Dresden, ME 75
Dummerston, VT 19, 34, 79, 80, 90–92
Dupont Company 142
Durham, ME 37, 57, 75, 145
Durham, NH 44
Durham-Lisbon Bridge (ME) 38, **145**–146

Eads Bridge (MO) 12, 41
East Machias, ME 56
East Rochester, NH 68
Eastern Railroad 44, 70; *see also* Maine Eastern Railroad
Edgecombe, ME 70
Edmundston, Canada 165
Edwards, Llewellyn 17, 34, 56–57, 64, 166
Elder, David 163
Elgin, IL 130
Ellet, Charles, Jr. 39, 163
Elm Street Bridge (NH) **140**
Elm Street Bridge (VT) 30, 36, 37, 83, 97–**98**
Emerson, Jim 174
Enosburg, VT 74
Erie Canal 10
Errol, NH 60
Evans, Hattie 135
Evans, Loring 135
Exeter, NH 19, 32, 67, 113

Fabyan House Hotel 137
Fabyan Station Bridge (NH) 67, 69, **137**–138
Fairfield, ME 13, **28**, 35
Fairlee, VT 23, 27, 181, 182
Falmouth, ME 66
Fay, Spofford, & Thorndike Co. 63, 117–118
Federal Aid and Works Program 15
Federal Bridge (NH) 52, 53, 54, 57
Federal Highway Administration 15
Fifteen Miles Falls Development Project 176–177

Index

Finley, James 10, 39
First Iron Bridge (NH) 134
Firth of Tay Bridge (Scotland) 10
Fisherville Bridge (NH) 53, 54
Fletcher, Bela 8, 182
Flood of 1927 73–75, 92, 101, 103, 105, 180
Flood of 1936 75–76, 119, 144, 152, 161, 182, 184
Florianapolis Bridge (Brazil) 63
Fore River Bridge (ME) 44
Fort Kent, ME 37, 50, 51, 166
Foss, Roland 163
Foundry Road Bridge (VT) 34, *102*
Fourth Iron Bridge (NH) 68, 132, 134
Fox River Bridge (IL) 130
Franconia, NH 19, 35, 48, 80, 130–32
Frank J. Wood Bridge (ME) 144
Frankenstein, Godfrey 136
Frankenstein Trestle (NH) 65, 135–*136*
Franklin, NH 24
Free-Black Bridge (ME) 36–37, *144*–145
Freedom, NH 59
Fremont, NH 30
Fuller, Levi 18

Gale River 130, 131
Gambo Falls Bridge (ME) 82, 86, 141–*142*
Garvin, James 84–85
General Sullivan Bridge (NH) 38, 63, *117*–118
Georgetown, ME 148
Gilead, ME 64
Glen, NH 36
Glimmer Glass Bridge (NJ) 154
Goffstown, NH 13
Golden Gate Bridge (CA) 40, 112, 157, 163
Good Roads League (NH) 16–17
Good Roads Movement 14–17
Gorham, ME 141
Gorham, NH 40, 75
Gorham Pewter Company 168
Grand Trunk Railroad 33, 56, 64, 67
Granite Street Bridge (NH) 35
Granite Street Bridge (VT) 36, 80, 103–*104*
Granville, VT 61
Graves, Edwin 159
Graves, Rufus 6
Great Arch Bridge (NH) 6, 117
Green Mountain Flyer 71
Green Mountain Railroad 71, 93
Greenfield, NH 25, 82, 114
Gristmill Bridge 72, *73*
Gronquist, Carl 157
Groton Bridge Company 26, 48, 114, 138, 162
Guildhall, VT 174
Guilford Transportation Corporation 132
Gulf Bridge (NH-VT) 185; *see also* Chief Justice Harlan Fiske Stone Bridge
Gulf Stream Trestle (ME) 65–*66*

Hagen & Thibodeau Construction Company 17, 182
Ham Branch 130
Hamilton Bridge Company 60
Hampden River Railroad Bridge (ME) 56
Hampton, NH 44
Hampton River Toll Bridge 44
Hancock, NH 17, 25, 82, 114
Hancock-Greenfield Covered Bridge 59
Hardesty & Hanover Company 62
Hare, Sheldon T. 54, 60
Harmony, ME 81
Harrington & Cortelyou Company 62, 171
Harrington, John L. 45, 62, 171
Harris, D.L. 60
Harrison, ME 25, 30, 34, 80, 86, 159, 160
Hartford, VT 13, 42, 58, 68, 74, 96, 100
Hart's Location, NH 36, 68, 70, 132, 135
Haverhill, NH 70, 177, 178
Hawkins, Richard F. 60, 61, 130
R.F. Hawkins Iron Works 60, 95, 114
Hawthorne Avenue Bridge (OR) 168
Henderson, Cornelius Langston 166
Henniker, NH 25, 30, 54, 58, 124
Henry Ford Museum 121
heritage railroads 70–71
The High Bridge (NH) 65
Highgate Falls, VT 13, 19, 28, 30, 35, 47, 50, 74, 106, 107
Highgate Falls Bridge 19, *20*, *29*, 50, 106–*107*
Hilton truss 34, 90–91
Hinesburg, VT 81, 86
Hinsdale, NH 36, 58
Hobo Railroad 71, 126
Holderness, NH 127
Hollis, ME 38, 57
Holt Road Bridge (VT) 99–*100*
Hooksett, NH 75–76, 118
Hooksett Railroad Bridge *76*
Hool, George 59
Houlton, ME 35
Howard, Marilyn Morse 90
Howard Hill Bridge (VT) 19, *33*, 86, 95
Howe, William 8, 60
Howe Bridge Works 60
Howe truss 11–12, 45
Howland, ME 38
Huntress Road Bridge (NH) 59
Hutchinson, John 34

I-35W Bridge (MN) 79
I-91 Bridge (VT) 39
I-95 Bridge (NH-ME) *171*
indeterminate structures 20
Industrial Revolution 13

International Bridge (Fort Kent, ME) 50, *51*
International Bridge (Madawaska, ME) 57, *165*–166
International Bridge (Van Buren, ME) 17, 50
Interstate Bridge (NH-VT) 35
Interstate Bridge Commission (NH-ME) 170
The Iron Bridge (VT) *70*

Jackson, NH 37, 83
Jackson & Moreland Company 59
Jamaica, VT 92
Janice Peaslee Bridge (NH-VT) *173*–175
Jarvis, Charles 35
Jay, ME 75
Jay Corners, VT 35
Jefferson, NH 75
Jennings, E.B. 60
Jewell, Gov. Marshall 15
Johnson, John 6, 8
Johnson, VT 13, 74
Jones & Laughlin Steel Company 115
Jones Crossing Bridge (NH) 82, 112–*113*
J.R. Worcester & Company 63, 177

Kennebec River 146, 147, 151, 152
Kennebec River Bridge (ME) 6, 45, *151*–152
Keyes Bridge (VT) 50, 106
Keystone Bridge Company 65
Kimberton, PA 121
King, Harry 61
King, James 61
King, Zenas 61
King Iron Bridge Company 33, 48, 49, 61
king-post truss 31
Kingfield, ME 164
Kinne, W.S. 59
Kittery, ME 26, 46, 62, 72, 79, 167, 168, 170, 171
Kittredge Bridge Company 173
Knight & Crosby Bobbin Factory 128

Lackawanna Steel Corporation 54, 118, 149
Lahood, Secretary of Transportation Raymond 169
Lake Champlain Bridge (VT-NY) 27, *38*, 72, 79, 118
Lake Champlain Flyer 71
Lancaster, Col. Thomas 8
Landry, Robert 27
Langdon Street Bridge (VT) 83
Langley, Harold E. 59, 173, 182, 184
League of American Wheelsmen (Maine chapter) 17
Learnard, Earnest 146
Leeds, ME 37, 145
lenticular truss 34–35, 106, 107, 116, 126, 130
Letter "S" Trestle (ME) 65
Lewis, Duane 150–*151*

Lewis, Dwight 150
Lewis, Mildred 151
Lewis, Norman 150–151
Lewis, Roy 151
Lewiston, ME 75
Lilac Bridge (NH) 76, 118–*119*
Limington, ME 73, 75
Lincoln, NH 71, 126
Lisbon, ME 38, 75, 145
Little, Moses 127
Little Bay Bridge (NH) 118
Little Harbor Bridge (NH) 44
Littleton, NH 59, 75, 177
Livermore, Judge Arthur 127
Livermore Falls Bridge (NH) 19, 35, *126*–128
Londonderry, NH 60
Long, Col. Stephen 7–8, 22
Lothrop Bridge (ME) 31
Lottery Bridge (NH) 33
Ludlow, VT 74, 93
Lund, Charles C. 52–53, 58
Lunenberg, VT 175
Lyman Bridge (NH-VT) 176
Lynch, Gov. John 185
Lyon Granite Company 91

Machiasport, ME 56
Mackinac Strait Bridge (MI) 63
Madawaska, ME 37, 165
Maidstone, VT 19, 33, 115, 173
Main Street Bridge (NH) 53
Maine Central Railroad 64–66, 68, 70, 75, 132, 134, 137, 141, 145, 146, 153
Maine Eastern Railroad 71, 147, 153
Maine Historic Preservation Commission 85, 86
Maine Kennebec Bridge Company 152
Mainely Brews Restaurant and Brewhouse 159
Manchester, NH 13, 14, 15, 23, 32, 35, 75
Manchester & Keene Railroad 67
Manchester Street Bridge (NH) 54
Mann, William 58
Marine Parkway Bridge (NY) 45
Marsh, Sylvanus 137
Martin, Harry 54
Mason Street Bridge (NH) 35
Mattawankeag, ME 66
Max L. Wilder 17
Max L. Wilder Memorial Bridge (ME) 39, *148*
McCallum truss 124
McClintic-Marshall Company 146, 182
McGregor Bridge (NH) 14, 15, 75
McIndoe Falls Bridge (NH-VT) *176*–177
McIntosh, Herbert 61
Meadows Bridge (NH) 24, 28, 47, 79, 81, *138*–*139*
Mechanic Falls, ME 13, *14*
Medburyville Bridge (VT) 34, 83, 89–*90*
Megquier & Jones Company 144

Memorial Bridge (NH-ME) 12, 26, 46, 72, 79, 85, 167–*168*, 169–*170*
Meredith, NH 71
Merrimack River 118, 119, 123, 124
Merritt, Chapman, & Scott Company 155, 157
Middlebury, VT 69
Milan, NH 72
Milford, NH 58, 111, 112
Mill Street Bridge (VT) 26, 37, 83, 98–*99*
Missisquoi River 106, 107
Mitchell, Moses 164
Monadnock Mills Footbridge (NH) *121*–122
Monadnock Mills #6 26, 32, 121–122
Monroe, NH 176
Montpelier, VT 13, 36, 70, 74, 77, 80, 83, 84, 85, 86, 103, 104
Montpelier & Wells River Railroad 68, 97
Montreal Express 68
Montreal, Maine, & Atlantic Railroad 67
Monzani, Willoughby 34
Moretown, VT 74
Morgan, J.P. 55, 61, 95
Morristown, VT 74
Morse, Col. F.B. 163
Morse, Harry 90
Morse, Silas 139
Morse, Viola Bishop 90
Morse Bridge (ME) *160*–*161*
Morse's Fool Bridge (ME) 164
Moseley, Thomas 61, 121
Moseley Iron Bridge Works 13, 32, 121
Mount Hope Bridge (RI) 63
Mount Orne Bridge (NH-VT) 17
Mount Washington Cog Railway 137
multiple intersection truss 34
Murphy Road Bridge (VT) 32, 81

Nashua, NH 13, 35, 110, 111
Nashua Manufacturing Company Bridge *110*–111
Nashua River 110
Nashua River Watershed Association 71
National Bridge & Iron Works 37, 98
Neil R. Underwood Memorial Bridge (NH) 44
Nelson, Kathryn 71
New Boston Railroad 58
New Castle, NH 44
New Deal Administration 15, 146, 172
New England Bridge Works 61
New England Central Railroad 102
New England Power Company 176
New Hampshire Bureau of Bridge Design 27
New Hampshire Department of Historic Resources 85–86
New Hampshire Toll Bridge Commission 117

New Haven, VT 93
New Portland, ME 162, 163, 164
New Sharon, ME 25, 37, 161, 162
New Sharon Bridge (ME) 161–*162*
Newbury, VT 177, 178
Newburyport Bridge (MA) 10
Newfane, VT 74
Newfields, NH 45
Newington, NH 38
Newington–Dover Point Bridge (NH) 45
Newman, Scott 108
Newport, NH 32, 60
Niagara Bridge Works 65, 135
North Andover, MA 121
North Conway, NH 134
North Hero, VT 44, 60
North Hero Bridge (VT) 60
North Monroe, NH 177
Northfield, VT 19, 37, 83, 86, 102, 103
Northwoods Stewardship Center 175
Norton, Henry 95
Norwich University 61
Notre Dame Bridge (NH) 62

Oak Street Bridge (NH) 23, 60
Oakland, ME 66
Office of Public Roads 15, 56
Old Orchard Beach, ME 65
Old Woodsville–Wells River Bridge (NH-VT) *178*–179
Ordway, S.S. 106
Orford, NH 23, 27, 59, 181, 184
Oriental Powder Company 141–142
Orono, ME 159
Osgood Bridge (NH) 27, 61
Ottauquechee River 96, 99, 100
Otter Creek 105

Paddock Road Bridge (VT) 26, 36, 80, 93–*94*, 100
Paine, Thomas 10
Palmer, Timothy 6, 7
Paradis, Bill 174–175
Park Street Bridge (NH) 19, *113*–114
Parker, Charles H. 11, 37, 98
Parker, Samuel 164
Parker patent truss 37, 97–98
Parker pony truss *103*
Parker truss 37, 94, 124, 151, 152
Patterson Hill Bridge (NH) 25, *26*, 30
Pauli truss 35
Pawlet, VT 85
Peavine Bridge (VT) 100–*101*
"Peavine" Railroad 68, 101
Pembroke Bridge 53, 54
Pemigewasset River 125, 126, 127
Pemigewasset River Railroad Bridge *125*–126
Penn Bridge Company 53
Pennsylvania Railroad 32, 36
Pennsylvania Steel Company 68, 132, 145
Pennsylvania truss 36, 98, 104, 105,

Index

108, 107, 109, 152, 153, 161, 166, 180, 182
Penobscot Narrows Bridge (ME) 158
Penobscot River 156, 157
Pettit's triangular truss 36
Philbrick, Augustus 138
Phillips, ME 67
Phoenix Iron Bridge Company 20, 45, 72, 120, 140, 155, 171
Piermont, NH 36, 75, 180
Piermont Bridge (NH-VT) **180**
Piermont Bridge Company 180
Pingree Bridge (NH) 19, 25, 30, 79, 81
Pioneer Street Bridge (VT) 86, 104
Piper, Walter 8
Piscataqua River 167, 169, 170, 171
Piscataquis River Bridge (ME) 37
Pittsburgh Bridge Company 61, 67
Pittsburgh–Des Moines Steel Company 161
Pittsfield, ME 32
Plymouth, NH 127
Plymouth & Lincoln Railroad 36
pony truss 23, 114, 115
Portland, ME 19, 32, 36, 44, 64, 69, 140
Portland & Ogdensburg Railroad 65, 135
Portland & Rumford Railroad 67
The Portland Company 64
Portsmouth, NH 26, 46, 62, 72, 79, 167, 168, 170, 171
Portsmouth, Dover, & York Street Railway 169
Portsmouth Naval Shipyard (NH) 171
Poultney, VT 85
Powers, Nicholas 8
Pratt, Caleb 11
Pratt, Henry, Jr. 17, 58
Pratt, Thomas 11, 32
Pratt truss 32, 33, 36, 115
Presumpscott River 141–142
Prospect, ME 156
Prowse, Robert J. 60
Public Service Company of New Hampshire 137
Public Works Administration 77, 146, 154, 172
Pumpkin Seed Bridge 127; *see also* Livermore Falls Bridge

Quichee Gorge Bridge (VT) 58, 68, **96**–97, 100

Rabbitt Hollow Bridge (VT) 37, 102–**103**
Rainbow Bridge (NH) 124
Ranger Bridge (NH-VT) 178
Rattlin' Bridge (VT) 105–**106**
The Red Bridge (ME) 25
Reed, George 61
R.F. Hawkins Iron Works 60, 95, 114
Rice Farm Road Bridge (VT) 19, 34, 79, 80, **91**–92

Richford, VT 35 74
Richmond, ME 75, 76, 85, 149, 151
Richmond, VT 36, 45, 57, 74, 105
River Street Bridge (VT) 24
Robie, George 119
Robie's Country Store 119
Robinson, Holton 154, 156, 157
Roby, Luther 53
Rochester, VT 68, 100
Rochester Shoe Tree Company 126
Rock River Bridge (IL) 130
Rockingham, VT 69, 75
Rockland, ME 71
Roebling, John 143, 163
Roleau, D.A. 106
Roleau, John 106
Rollins, Frank 16
Rollinsford, NH 23
Rouse's Point Bridge (VT) 44, 63
Route 175 Bridge (NH) 61
Royal Albert Bridge (UK) 35
Royalton, VT 74
Rumford, ME 75, 160, 161
Rupert, VT 85
Rural Free Delivery Act 15
Rutland, VT 13, 74
Rutland-Canadian Railroad 44, 58
Rutland Railroad 93
Ryan, Sally 174
Rye, NH 44
Ryefield Bridge (ME) 25, 30, 34, 80, 86, 159–**160**

Saco River 132
Saco River Bridge (NH) **28**, 30
Sagadahoc Bridge (ME) 147
St. Albans, ME 31, 33, 48, 81
St. Albans, VT 60, 61
St. Albans Iron and Steel Works 60
St. George River 153, 154
St. John River 165, 166
St. John Street Underpass (ME) 19, 36, 69, 140–**141**
St. John's Bridge (OR) 63, 155
St. Johnsbury, VT 132
St. Johnsbury & Lake Champlain Railroad 68
St. Johnsbury & Lamoille County Railroad 108
Salisbury, NH 19, 25, 30, 79, 81
Samuel Morey Memorial Bridge 23, 27, 62, **181**–182, 184
Sanders Hill Road Bridge (NH) 129
Sandy River 67, 161, 164
Sandy River & Rangeley Lakes Railroad 67
Sandy River Railroad Bridge (ME) **67**
Sarah Mildred Long Bridge (NH-ME) 46, 170–**171**, 172
Sargent, Hubert 61
Sargent, Paul D. 17, 56
Sassanoa River 148
Sassanoa River Bridge 148; *see also* Max L. Wilder Memorial Bridge
Sawyers Bridge 25
Sawyers River 132
Saxton's River, VT 35

Scarboro, ME 65
School Street Bridge (VT) 83
Scribner Road Bridge (NH) **30**
Seabee Bridge (NH-VT) 185; *see also* Chief Justice Harlan Fiske Stone Bridge
Sebago Lake 142
Second Iron Bridge (NH) 69, 132–**133**, 134
Sedgewick, ME 154
Sewall's Falls Bridge (NH) 24, 53, 79, **123**–124
Shaftsbury, VT 31
Sharon, VT 74
Sheepscott River 152
Sheepscott River Railroad Bridge (ME) 66, **152**–153
Sheffield Scientific School (Yale University) 35
Shelburne, NH 24, 28, 47, 75, 79, 81
Sheldon, VT 74
Shoreham, VT 21
Simpson Brothers Construction Corp. 124
Skowhegan, ME 157
Smith, C.B. 102
Smith, Royal 102
Smith, Wallace 102
Smithsonian Institute 32, 57
Snowe, Senator Olympia 162
Snowman, Ruth Lewis 151
Solid Lever Bridge Company 37
Somerset Railroad 65, 66
Souhegan River 111, 112
South Halstead Street Bridge (IL) 45, 168
South Paris, ME 33
South Wallingford, VT 31
Southport, ME 45, 148
Southport Bridge (ME) **148–149**, 150–151
Spofford, Charles M. 63, 117
Springfield, MA 95
Springfield, VT 13, 26, 36, 37, 70, 80, 93–94, 100, 182, 183
Springfield Electric Railway 70, 94, 182
Springfield Trails & Greenways 94
Squamscott River Bridge (NH) 45
Stanley, Edwin 35
Steele Road Bridge (NH) 81
Steep Falls Bridge (ME) 73
Steinman, David B. 55, 63, 85, 154–157
Steubenville Bridge (OH) 63
Stewartstown, NH 75, 172
The Stimulus 84–85, 169
Stinchfield Bridge (ME) 37
Stockbridge, VT 74
Stone, Amasa 8, 11, 32, 60
Stone, Chief Justice Harlan Fiske 185
Storrs, Edward 54, 58, 97, 123
Storrs, John W. 16, 54, 57–58, 68, 70, 79, 97, 113, 119, 123, 124, 125, 128, 178, 180

Stratford, NH 19, 33, 35, 64, 115, 173–174
Stratford Hollow Bridge (NH-VT) 173; see also Janice Peaslee Bridge
Stratford Hollow–Maidstone Bridge Association 174
Stratham, NH 45
Strong, ME 67, 164
Sugar Hill, NH 75, 80
Sugar River 121, 122
Sullivan Machine Company Bridge (NH) 36, 70, *122*–123
Sulphite Railroad Bridge (NH) 64
Suncook River 116
Suncook Valley Railroad 67
Sunderland, VT 31
Swanton, VT 24, 26, 44, 81, 86, 107, 108
Swanton Footbridge 107, *108*–109
Swanton Historical Society 108
Swanton Village Electric Company 107
The Swing Bridge (ME) *143*–144
The Swinging Bridge (NH) *111*–112

Tacoma Narrows Bridge (WA) 156
Tamworth, NH 54
Tannery Bridge (ME) 31–32, *33*, 48, 81
Tasker, James 8, 182
Taylor Street Bridge (VT) *84*–85
Taylor's Falls Bridge (NH) 13, 35
Thetford, VT 86
Third Iron Bridge (NH) 68, 132–*133*, 134
Thomaston, ME 44, 153, 154
Thompson Crossing Bridge (NH) 19, 33, 82, 115–*116*
Thornton, NH 36, 81
Thousand Islands Bridge (NY) 154
Three Bridges Crossing (VT) 68
Thunder Bridge (NH) 25, 28–*29*, 35, 82, 86, *116*–117
Ticonic Falls Railroad Bridge (ME) 66
Ticonic Footbridge Company 159
Tilton, Charles 129
Tilton, Nathaniel 129
Tilton, NH 24, 25, 26, 64, 129, 130
Tilton Island Park Bridge (NH) 25, 26, *129*–130
Toonerville Trail 94
Topsham, ME 33, 65, 76, 142, 143, 144
Topsham Railroad Bridge (ME) *76*
Town, Ithiel 7–8
Towne, John 34
Townsend Gut Bridge (ME) 45, 148, 150; see also Southport Bridge
TransCanada Corporation 177
Tropical Storm Irene 77
Truesdell, Lucius 130
Truesdell truss 11, 26

Tunbridge, VT 34, 102
Turk, J.C. 67
Turner, Charles 62
Turner, ME 75
The Twin Bridges (NH) 53
Twin Mountain, NH 75
Two Cent Bridge (ME) *158*–159

Union Bridge (NH) *21*
Union Bridge Company 136
United Construction Company 124, 160
United States Steel Corporation 55
University of Maine 61

Varnum, H.W. 106
Vermont Construction Company 49, 60–61, 95, 102
Vermont Historic Bridge Program 33, 81, 86, 107
Vermont League of Good Roads 18
Vermont Railway 93
vernacular structures 30
Verona, ME 156
Verona Island Bridge (ME) 157
Vierendeel, Arthur 157
Vierendeel truss 157, 158
Vine Street Bridge (VT) 19, 37, 83
von Pauli, Friedrich Augustus 35
Vose, George 48–49, 55–56
Votey, Josiah 61

Waddell, John Alexander Low 26, 45, 62, 146–147, 168, 169, 171
Wadsworth Street Bridge (ME) 44, *153*–154
Waldo-Hancock Bridge (ME) 63, 79, 154, 155–*156*, 157–158, 165
Wallingford, VT 70, 86, 93
Walpole–Bellows Falls Bridge (NH-VT) 62
Wardsboro Brook 92
Wardsboro Brook Bridge *92*
Warren, James 34
Warren, VT 74
Warren truss 33–34, 105; multiple-intersection 100, 101, 137, 159
Washington County Railroad 83
Waterbury, VT 74
Waterville, ME 66, 75, 158
Watson Bridge (NH) 25
Watson Road Bridge (NH) 80, 81
Weatherby, Farewell 8
Wells, John H. 59, 182, 184
Wells River, VT 36
Wentworth, NH 128–129
Wentworth Village Bridge (NH) *128*–129
West Buxton Bridge (ME) 38
West Dummerston Bridge (VT) 91
West Milton, VT 26, 36, 36, 81, 108
West River Bridge (VT) 18, 90
West Rutland, VT 86
Western Avenue Bridge (NH) 124

Westfield, VT 86
Westport, NH 68
Weybridge, VT 105, 106
Wheeling Suspension Bridge (OH-WV) 163
Whipple, Squire 10–11, 32, 33, 34, 37, 45
Whipple truss 28, 33–34, 65
Whitcher's Falls Bridge (NH) 25
Whitcomb, Col. Benjamin 175
Whitcomb Bridge (NH-VT) *175*–176
White Mountain School 136
White River Bridge (VT) 68
White River, First Branch 102
White River Railroad 68
White River, Third Branch 100
White River Valley Railroad 100
Whitney, Miss A. 139
Whitneyville, ME 56
Wilcox, S.C. 35
Wilder, Max L. 57
Willey, Samuel 135
Willey Brook Bridge 68, 70, *134*–135
Wilmington, VT 34, 77, 82, 89–90
Wilson, Ezra 164
Wilton, NH 32
Windham, ME 13, 82, 86, 141, 142
Winnepesaukee River 129
Winnepesaukee Scenic Railroad 71
Winnesquam Bridge (NH) 24
Winooski River 103, 104
Winslow, ME 159
The Wire Bridge (ME) 162–*163*, *164*–165
Wiscassett, ME 66, 70, 152
Witham, William 164
Wolcott, VT 74
Woods, Dutton 53
Woodstock, NH 36
Woodstock, VT 26, 30, 35, 36, 37, 61, 68, 83, 97, 99
Woodsville, NH 36
Woodsville–Wells River Bridge (NH-VT) *177*–178
Woolwich, ME 46, 146, 148
Worcester, Joseph R. 62, 178
Worcester, Thomas 62
J. R. Worcester & Company 63, 177
Works Progress Administration 77, 117, 124, 144
Worumbo Mill Company 146
wrought iron 11–12
Wrought Iron Bridge Company 32, 48, 52, 53, 113
Wynkoop, James 53

Yarmouth, ME 13, 35

Zenas King Memorial Bridge (VT) 61
Zoller, Jerry 27

www.ingramcontent.com/pod-product-compliance
Lightning Source LLC
Chambersburg PA
CBHW080936020526
44116CB00034B/2897